HOW TO RAISE
PIGS
EVERYTHING YOU NEED TO KNOW

PHILIP HASHEIDER

Voyageur Press

DEDICATION

This book is dedicated to my late nephew, David Hasheider, whose deep passion for pigs and enthusiastic and lighthearted personality was an inspiration for many 4-H and FFA members across Wisconsin.

First published in 2008 by Voyageur Press, an imprint of MBI Publishing Company, 400 First Avenue North, Suite 400, Minneapolis, MN 55401 USA

Voyageur Press titles are also available at discounts in bulk quantity for industrial or sales-promotional use. For details write to Special Sales Manager at MBI Publishing Company, 400 First Avenue North, Suite 400, Minneapolis, MN 55401 USA.

To find out more about our books, visit us online at www.voyageurpress.com.

Library of Congress Cataloging-in-Publication Data

Hasheider, Philip, 1951-
  How to raise pigs : everything you need to know / by Philip Hasheider. -- Rev. and updated.
    p. cm.

  Includes index.
  ISBN 978-0-7603-4525-2 (softcover)
  1. Swine. I. Title.
  SF395.H745 2014
  636.4--dc23
                                    2013021906

Editors: Amy Glaser, Elizabeth Noll
Design manager: James Kegley
Cover design: Carol Holtz
Interior design: Mandy Kimlinger
Layout: Kazuko Collins

Photo credits
Front cover: Colin Edwards / Alamy

Printed in China

10 9 8 7 6 5 4 3 2 1

CONTENTS

INTRODUCTION
Pigs and People 4

CHAPTER 1
Breeds of Pigs 12

CHAPTER 2
Getting Started 30

CHAPTER 3
Housing and Fencing 54

CHAPTER 4
Feed and Manure 78

CHAPTER 5
Breeding and Reproduction 98

CHAPTER 6
Managing and Showing 110

CHAPTER 7
Butchering and Marketing 128

CHAPTER 8
Country Living, Getting Help, and Exit Strategies 160

APPENDIX 1
Glossary 170

APPENDIX 2
Resources 172

INDEX 174

ABOUT THE AUTHOR 176

ACKNOWLEDGMENTS 176

INTRODUCTION
PIGS AND PEOPLE

Pig raising has changed dramatically over the years, as food preferences and customs have changed. Pig lard used to be a household staple, and a fat pig was considered ideal. These days lean pork has wide consumer appeal. The old style versus the new style of pigs has led to major changes in how they are raised. There are also many different reasons to raise pigs: as food, for byproducts, for medical applications, and as pets. And there are different sizes of pig-raising programs, from a small family-owned herd to a large commercial unit.

Raising pigs can be a young person's pathway into 4-H or FFA (formerly Future Farmers of America) livestock programs, since pigs require little investment and can become saleable in about five to six months' time. For others, pig raising can provide a source of food for the family, as well as a sense of satisfaction that comes from working with one of nature's most intelligent animals.

Pigs do not require extensive housing or equipment to be raised in a humane and safe manner. With a little guidance and sufficient knowledge, any fledgling pig grower can successfully raise pigs from birth to market. Pigs are among the easiest animals to raise if you're new to the countryside and wish to embark on a small livestock-rearing enterprise. Pigs, with their manageable size, ability to consume feedstuffs other animals won't eat, general nature, innate intelligence, and low initial investment, are ideally suited for small-scale farms.

Why do you want to raise pigs? Reasons may range from a homegrown food source for your family to a pleasurable pastime and the aesthetic value of having unique personalities on the farm. Pigs were once known as "mortgage lifters," and such as endeared themselves to many small-scale farmers.

Raising pigs has several advantages when compared to raising other species of livestock, and most of these advantages relate to the investment of money and time. By the time a weaning pig is a month old, it can be fairly self-sufficient. It can find feed and water by itself with little assistance and be off to market or butchered in less than a year. Pigs can utilize virtually any edible feed source and grow quickly, which makes it possible to raise two litters of pigs from one sow in any given year.

In contrast, a beef calf can require twelve to fifteen months of care before it reaches market weight. A dairy calf is still an infant for many months and needs special feed, such as milk or milk replacements, to ensure sufficient growth until its rumen develops to where it can eat solid feed. A dairy cow requires considerably more care and feed, plus facilities and equipment for a milking system.

With more than eight major pig breeds and several heritage breeds available in the United States, you should be able to choose a breed of swine with characteristics that best fit your farm. Breeds differ in their meat-producing ability, mothering instincts, and reproductive performance. You can adopt the breed that best fits your management system.

PIGS FOR PROFIT

The spectrum of conditions under which pigs are or can be raised may be as varied as the farms on which they live. The term "mortgage-lifters" came about because, with little outside feed purchased, farmers could consistently produce a marketable animal that could offset many expenses. Farmers could raise enough hogs each year to pay for the farm mortgage. In the mid-1960s, a shift in specialization on farms occurred. Farmers moved away from raising many species of livestock, such as chickens, beef, dairy, swine, and sheep, and focused only one or two kinds of animals.

Other forms of agriculture also evolved into what is known as agribusiness. Farms became larger, and farmers devoted more time and expertise to fewer species of animals. Similarly, as price fluctuations and profit margins altered different livestock markets, specialization in one or two species became the norm. There was also a noticeable shift in the number of people living on farms or wanting to work on them, and labor issues started to emerge.

The trend toward livestock specialization—such as raising solely dairy, beef, or swine—accelerated during the 1970s and 1980s to the point where today few large farms raise multiple livestock species. If a farming enterprise does own more than one species, it typically has separate units on separate farms, and the animals do not commingle.

Small-scale pork production can still be a viable option if you're willing to seek out niche markets. Pigs can be raised outdoors in fields or pastures where they can roam in all kinds of weather and thrive. Pigs are also found in large commercial operations where hundreds or thousands of animals can be housed, fed, and raised in close confinement units that are highly protected bio-security areas.

Raising pigs offers a variety of options because of the very nature of the animal. You can enter and exit a small-scale pig-raising enterprise quickly, with only a short-term commitment needed from birth to market. With minimal investment, you can establish a small herd of one or two sows, raise their litters to market weight, sell the pigs, and then exit within seven to eight months. Then, if you wish to return at a later date, you can re-enter the pig production business with the purchase of pregnant sows.

While it is extremely difficult to make any kind of substantial profit on two or three sows and their resulting litters, this likely is not the sole reason for you to raise pigs. If profit is your sole motive, then you should consider a large scale enterprise that finishes out a large number of hogs in a short time period.

The beginning of a pig-raising program can be the dawn of a new life for you and your family. Many opportunities await if you choose to move to the countryside and become part of a farming community.

This is not to say that you can't make money raising pigs on a small scale. You can, but the market for the animals you raise will need to be one that attracts customers who appreciate the product and the way in which it has been raised.

PIGS FOR PLEASURE

Pigs can easily be raised on your farm for personal enjoyment. If profit is not your sole intention for raising pigs, interacting with a somewhat large but usually gentle animal can have aesthetic rewards beyond other livestock. They can become pets.

Pigs are creatures of habit and attune themselves to their surroundings and treatment like other livestock. The more you work with and handle your pigs, especially if you raised them from piglets, the more they will acclimate themselves to their environment and respond to your treatment. They will quickly become familiar with your voice, mannerisms, and animal sense.

A person who raises pigs can derive a deep satisfaction from working with an intelligent animal and enter a world where a level of communication and understanding exists between animals and humans. Pigs can provide an avenue for young people to become involved with 4-H and FFA programs in local, state, and national levels by exhibiting at fairs or carcass classes. These programs can influence career choices and offer experiences that enhance the life skills of youth.

PIGS WITH PERSONALITY

Perhaps no other farm animal has had the notoriety that pigs have had through history. Cows may jump over the moon. Chicken Little may have thought the sky was falling. Old Dobbin might make his annual Christmas appearance hitched to his sleigh. But for sheer variety and subject matter, pigs can claim to be the subject of the most varied and easily recognizable characters in many forums. The world of folklore, literature, film, and storytelling, not to mention nursery rhymes, is full of characters and personalities based on pigs.

In the days of exploring the high seas, pigs were kept on sailing vessels because many sea captains believed the pigs would swim toward the nearest available land if they were shipwrecked. In England, peppermint pigs symbolize the Victorian belief in the pig as a good luck charm. German well-wishers bring marzipan pigs to people for good luck at Christmas and New Year's. In the Chinese art of positioning objects, feng shui, golden pigs are believed to bring great prosperity and happiness to a household. In Chinese mythology, the pig is a symbol of honesty, tolerance, initiative, and diligence. In many cultures, children are traditionally given piggy banks to encourage them to save money.

In literature, Charles Dickens pointed out the value of pigs in *Great Expectations* when he wrote, "If you want a subject, look to pork!" Piglet, who figured so prominently in A. A. Milne's Winnie-the-Pooh adventures, took on human vulnerabilities, endearing him to children who could empathize with the innocent predicaments in which he found himself and rejoice at his successful emergence through his trials.

In *Charlotte's Web*, E. B. White used Wilbur, the runt of the litter, to befriend Charlotte, the spider. For better or worse, Snowball and Squealer were pigs who played a significant role in George

Before and during the 1940s, a large, heavy hog was desirable for the fat it produced that could be rendered into lard to be used as a food staple from cooking to canning. As consumer diets began to change and substitute vegetable products were introduced, lard was no longer in demand. Farmers had to respond by breeding and sculpting a leaner hog with a higher meat-to-fat ratio.

During the 1950s, a shift began to a leaner pig that was more in demand by the hog markets and consumers. The author's father appears here with one of his prize-winning registered Chester White boars that he used in his pig-breeding program.

point of sexual maturity, the term *hog* can be used to denote an animal that is not only approaching market weight but also at a stage for reproducing. In its broadest sense, the term *swine* can mean a herd of pigs.

Most of the swine breeds known today are widely thought to descend from the Eurasian wild boar, *Sus scrofa*, similar to today's wild hog. Pigs were among the earliest animals to be domesticated, and paintings and carvings of pigs, dating back over 25,000 years, have been found. There is fossil evidence dating back 40 million years that indicates wild pig-like animals roamed forests and swamps in Europe and Asia.

Archaeology in the Middle East and China has provided evidence of widespread domestication as early as 4900 B.C. and in Europe by 1500 B.C, suggesting that the pig was most often associated with settled farmers rather than nomads, perhaps because of the difficulty in herding pigs in open areas.

Several major world religions have not looked kindly on pigs or swine, as evidenced by some of the practices embodied in their teachings. Why pigs are prohibited within any religious dietary laws or customs has been debated for centuries and may be the result of early concerns for health and safety in consuming foods or liquids.

Orwell's *Animal Farm,* where some of the other pigs evolved from helpful, cooperative farm animals to being the task masters of the entire farm. Nursery rhymes such as "This Little Piggy" have become favorites of parents and are used to elicit giggles and laughs from toddlers to older children. Stories involving the adventures of the Three Little Pigs have delighted children for generations.

In the television sitcom *Green Acres,* Arnold Ziffel gained fame and notoriety as a pig/son. *Babe* is a book and a movie about a pig by that name who is raised by a border collie and rounds up the other animals, becomes a film star, and receiving a reprieve from being butchered by talking with the sheep and convincing them it is in their mutual interest to cooperate with him. Subsequently, Babe provides his owner with more than a prize ribbon. Finally, old sayings that contain elements of common sense and literal truth, such as "You can't make a silk purse out of a sow's ear," have entered the human vocabulary as proverbs.

A DIVERSE HISTORY

The porcine family of livestock has been commonly referred to as pigs, swine, or hogs. There really isn't a difference between these terms. In layman's terms, they mean essentially the same thing.

In a general way, however, the term *pig* can refer to young animals not sexually mature and weighing approximately 40 to 120 pounds. Prior to that, the term *piglet* can be readily applied. At the

Before the shift to intensive confinement in the 1970s and 1980s, pastures were typically a part of any pig-raising program. A shift is occurring once again, this time from confinement back to pasture grazing where possible. As more consumers request locally grown food products from animals raised in more traditional methods, small-scale producers have found a niche market for their pork products.

Raising pigs can be generational; interest can be passed from father and mother to son and daughter. Many farm families have benefited from working together to achieve common goals.
Pam Ziegler

Pigs are often thought to be dirty animals, particularly when seen wallowing in mud holes. This misconception has more to do with attitudes than actuality. Pigs generally keep themselves quite clean, evidenced by their status in some households as pets. Although pigs may be seen lying in mud, it is because they do not have sweat glands and constantly need water or mud to cool off on hot days.

PIGS ACCOMPANIED EXPLORERS
Although Christopher Columbus took eight pigs on his voyage to Cuba in 1493 at the insistence of Queen Isabella, it was Hernando de Soto who is typically dubbed the father of the American pork industry. He landed at what is now Tampa Bay,

Florida, with thirteen pigs in 1539. Reports note that within three years de Soto's herd grew to 700 head, not including the ones his troops had consumed, those that ran away and became wild pigs (the ancestors of today's feral pigs or razorbacks), and those given to Native Americans to keep the peace.

In 1607, Sir Walter Raleigh brought sows to Jamestown Colony, and during the ensuing century, pigs became a staple throughout the colonies. On Manhattan Island, residents constructed a long, solid wall on the northern edge of the colony to control free-roaming herds of pigs rampaging through their grain fields. The street that ended at the board wall was named Wall Street in what is now Lower Manhattan.

For generations, family farms have been the cornerstone of agriculture's success in the United States. You can play a part in its future success by becoming involved in this long tradition.

In past decades, a shift occurred when many rural people moved to villages and cities for a variety of reasons. In recent years this trend has slowly changed as some urban residents seek residence in the countryside. As farmland becomes available, it can provide a place for small-scale pig programs.

PIGS MOVE WEST

After the Revolutionary War, pioneers heading west brought their pigs with them, and as the western herds grew, the need for pork-processing facilities became apparent. Packing plants began to spring up in major cities as demand for pork grew and because of the large labor supply available to work in the plants. The first commercial slaughter facility was located in Cincinnati, Ohio, which soon became known as "Porkopolis," a place where more pork was processed than anywhere else in the Midwest.

The pork industry received a major boost soon after the Civil War when refrigerated railroad cars were introduced. This enabled pork processing to be centralized closer to points of production instead of near points of consumption. Major cities such as Chicago, Illinois; Kansas City and St. Joseph, Missouri; and Sioux City, Iowa, became terminal markets that were assisted by major rail lines crisscrossing the country through them. This enabled slaughtering plants to be located adjacent to the stockyards. Live animals could be shipped from great distances and held in these yards until they were sent through the slaughtering plants and processed for consumers. The finished meat products could then be shipped by refrigerated railcars to other distant points. This process became a revolving door and many farmers and packing houses benefited. Another shift in the industry followed as pork production relocated to the Midwest where supplies of feed grains and cereals were widespread. What was known as the "Corn Belt" became the "Hog Belt" as well.

TASTES FOR PORK CHANGE

Pigs eventually began to serve a dual role for the families raising them by supplying both meat and lard. Accelerated by the Industrial Revolution, urban populations grew. City dwellers used all sorts of fats and oils for their household needs, including cooking and making soaps and candles.

From a modern historical perspective, three eras in pork production stand out and explain as much about society's taste patterns as they do about farming trends. These can be described as the eras of lard, meat and bacon, and lean white meat. In 1832, a French chemist, Michael Chevreul, discovered how to synthesize liquid glycerin from lard and other solid fat compounds and use it to make inexpensive candles. This revolutionary process allowed people with limited financial means to afford candles to light their homes.

As demand for soaps, cooking fats, and candles increased, the price of lard increased, making pigs a valuable farm commodity. This continued until the arrival of petroleum products, vegetable oils, and electricity, which eventually provided cheaper alternatives.

The demand for lard affected the type of hog that was bred and raised. Meat became almost secondary as the lard-type hog became fashionable. This hog typically had short legs and a deep belly that consisted mostly of fat. Pictures of hogs reveal animals we would consider obese, but like other trends in society, diets change, and as these changes filter down through the markets, it alters the way farm animals are raised.

9

Farms supplied what the market demanded. Pigs of a similar structure and type proliferated across the countryside and eventually found their way into the show ring at fairs and livestock expositions.

As late as the 1950s, pigs were still being produced for the dual purpose of meat and lard. In recent years, as consumer diets have changed, the pig's main purpose has been to supply meat. The shift in consumer diets away from fats and lard to more meat was about as dramatic as the increased demand for lard decades earlier. Farmers responded by developing a leaner meat hog through breeding selection and new feeding programs.

INTRODUCING "THE OTHER WHITE MEAT"
A drop in pork consumption and demand in the early 1980s forced a change in pork composition. Between 1979 and 1985, pork demand fell 4 percent a year, while chicken sales dramatically increased.

This change in consumer diets sparked a transition within the pork industry that fostered new preferences and management practices. The pork industry developed a new marketing strategy and coined the slogan "The Other White Meat," which became one of America's five most recognized advertising slogans in the past twenty years.

With consumers concerned about fats in their diets, pork producers began raising animals that were 16 percent leaner than they had been fifteen years earlier and reduced the saturated fat content of the meat by 27 percent. While this shift marked

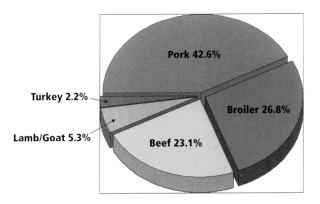

According to the United States Department of Agriculture, pork accounted for more than 43 percent of the meat consumed in the world in 2005. This chart represents the percentages in tons of pork, beef, and poultry. These three meat products account for 95 percent of the total meat consumed.

a significant alteration in the animal, it also caused the pendulum to swing too far. As the fat was removed, the meat's taste changed. This caused some consumers to move away from pork due to the lack of flavor.

Today the swing back to a more tasteful pork product has opened many opportunities for small-scale farmers to enter the swine business and develop niche markets for their pork products. Even with the alterations in the physical structure of pigs through time, pork continues to be the most popular meat worldwide.

BENEFITS TO HUMAN HEALTH
The wave of technology sweeping through society at an unprecedented rate in the last ten years has brought pigs to a place beyond that of a human food source. Nearly forty drugs and pharmaceuticals on the market come from pig sources. Today's technology is rapidly finding new uses for products derived from pigs. Pig heart valves have been transplanted into humans with consistent success. Pigs have become effective donors of brain cells, skin grafts (their hide is similar to human skin), insulin, and islets (endocrine cells that produce insulin).

As the medical profession continues to anticipate an increase in the development of Type II diabetes in large segments of the human population, pigs may become an important resource in its prevention and cure. At present, human islet cell transplants have reversed diabetes in 90 percent of recipients, but human donor organs are in short supply. Pig islets are in greater supply, and, because of the rapid growth of pigs, their tissue replication is more readily available for diabetes care.

Through human history the pig has provided an abundance of products that have enhanced people's lives. Ranging from food for their diets, lard for lighting their homes, soaps for cleaning, skin for grafting onto burn victims, valves to replace those in damaged hearts, and now with the prospect of helping cure diabetes, the value of the pig should not be underestimated.

EMERGING TRENDS AFFECT PIG RAISING
Energy issues continue to make news as the world market price of petroleum products has risen and fluctuated for any number of reasons. With the drive to develop renewable energy sources, one source being developed on a wider scale is that of ethanol-enhanced gasoline, as well as other products.

During much of the twentieth century many farms were stocked with several species of livestock that were raised with daily access to the outdoors. Separate areas on the farm were generally assigned to each species, and pigs in large lots had outdoor huts for shelter.

Ethanol production for use in gasoline engines for cars, trucks, and other motor vehicles requires a vast amount of corn. Whether sufficient amounts can be grown to meet this unexpected and large demand is still to be seen. While ethanol production also yields byproducts being tested and used in livestock feeds, product availability and purchase volume are factors that will impact their on-farm use.

The pig industry consumes approximately 10 percent of the corn crop in the United States. What does this have to do with raising pigs? The answer has both short- and long-term implications. In the short term, corn prices will likely rise overall and will impact expenses for those who purchase grain to feed pigs. This would dramatically squeeze the pig farmer's profit margin.

In the long term, this may be a benefit to a small-scale pig-raising enterprise. If you do not base your feeding program entirely on cereal grains and take advantage of the monogastric pig's versatility, the impact of rising corn prices may have a minimal effect on your profitability. By using other feedstuffs that may be readily available on your farm and developing your own market, you may be able to circumvent these higher costs and still maintain a degree of profit while raising pigs.

In other words, expected increases in corn prices will affect large enterprises, whether they specialize in pigs or cattle, in ways that may make them much less profitable or at least affect total production levels as farmers back away from feeding high levels of corn. This may become an opportunity for those who wish to raise pigs on a smaller scale to have both a profitable and satisfying swine operation.

CHAPTER 1
BREEDS OF PIGS

Before beginning your swine enterprise, consider what breed you might want to raise. Do you have any preferences? Is there a particular color or color pattern you like? Are your children interested in raising and showing pigs as a 4-H or FFA project?

You may wish to raise several breeds at the same time. This is easy to do and you may find this more interesting and challenging than raising just one breed. One of the drawbacks to raising several breeds, particularly if you wish to retain the breed strain or characteristics and not crossbreed them, is that you will have to keep a boar for each of those breeds. If you wish to keep it simple and maintain a herd of only one breed, the choices are fairly straightforward.

There are eight major swine breeds in the United States today that make up the vast majority of the registered and commercial pig population, and a number of minor or heritage breeds are gaining popularity among small-scale swine farms and those interested in preserving those breeds.

The emergence of artisan food markets has increased producer interest in supplying those markets with animals from less common breeds. Interest has increased in bringing back old-time flavors and textures to the dinner table. A "know-your-source" food ethic also can play a significant part in your pig-raising program and your choice of breed.

Some swine breeds had veered toward extinction before being raised and promoted as heirloom breeds. This has led to an animal conservationist movement, which includes sustainable food groups and like-minded chefs. By working with a heritage breed, you will join others who wish to restore vanishing breeds and the nation's livestock gene pool, which has been dramatically reduced by the breeding of animals for the mass market. Many heritage breeds can thrive on alternative feeding programs separate from fast-growth, high protein diets that push animals past their normal rate of growth.

Whether you decide to raise a heritage breed or acquire one that is considered more popular or mainstream, there are differences to consider. Knowing these differences will help give you more confidence in your choice.

CHOOSING A BREED FOR YOUR FARM

Each farm is unique, and the breed you choose may not be the one best-suited for someone else. Considerations, such as region of the country where you live, the climate, availability of a particular breed, and growth rates and mature size of the breed, may have to be taken into account.

All the purebred breeds offer a registry program. This involves pedigreed animals or purebreds versus unpedigreed or commercial animals. The parentage of purebred animals can be substantiated, which may allow them to participate in purebred programs and raise pigs for seed stock.

The purebred versus commercial markets do not need to be competitive as they only differ in approach and end goals. Raising quality animals can be accomplished whether or not they have a documented ancestry.

BREED VARIATIONS

Swine breed associations can provide information to help you select the breed appropriate for your goals and farm. Although individual animals may differ, some general guidelines and variations apply to each specific breed. Each breed, whether major or heritage, has unique characteristics that are to be appreciated for what they can help you accomplish with your program.

Deciding on which breed to raise can be an exciting time, and a discussion about the merits of each breed can involve the whole family. While the rare or less popular breeds may have their attractions for you, one important consideration is that no matter which breed you choose, it is likely any breed will do well if you pay constant attention to the pigs and provide good husbandry practices.

No matter which breed you raise, the pigs will need daily care or they will not perform in ways you will find acceptable. Their growth rate and well-being will be a reflection of the care they are given.

The eight major breeds are Landrace, Berkshire, Chester White, Duroc, Hampshire, Poland China, Spotted Poland China, and Yorkshire. Minor or heritage breeds include Hereford, Large Black, Mulefoot, Red Wattle, and Tamworth.

MAJOR BREEDS
LANDRACE

Landrace hogs were developed in Denmark at the end of the nineteenth century. The breed resulted from crossing the Large White hog from England with native Danish swine. In 1934, the United States Department of Agriculture received a shipment of the Danish Landrace, and many of those hogs were made available to agricultural experiment stations and used in crossbreeding programs.

The foundation Landrace stock were the pigs that were bred pure to the white strain or carried a small portion of Poland China blood, such as having a Poland China boar or sow in their third or fourth generation.

The modern breed has developed from this foundation plus an infusion of Norwegian, Danish, and Swedish Landrace blood, which was added through the importation of thirty-eight boars and gilts (young, unfarrowed female pigs) from Norway. Their blood was blended into the American Landrace to broaden the genetic base of the breed.

The Landrace is a white hog noted for its long body and having sixteen or seventeen pairs of ribs, which translates into more cuts of meat per animal. Because of this length, they generally do not exhibit the arch of back that other breeds typically have. In some cases, their backs are almost flat. Their droopy ears tend to be large, heavy, and carried close to the face. The females are known for being good mothers, producing large litters, and have milk-producing abilities to support the large litter.

Because of these qualities, the Landrace hogs are commonly used in maternal lines to incorporate these desirable characteristics. They are also popular because they are docile and easy to handle. Landrace sows do well either in confinement systems used during farrowing or in open spaces with plenty of room. Their pigs are noted for rapid growth and development.

The American Landrace originated in Denmark and has developed into a popular breed noted for its maternal line influence in many breeding programs throughout the country.
National Swine Registry

If you are raising Landrace pigs for registry it is important that the hair color is completely white. In an attempt to keep the breed pure, black hairs are not allowed and pigs that have black spots are not eligible for registration with the American Landrace Association.

BERKSHIRE

The legend surrounding the Berkshire is that the breed was discovered by the British army at its winter quarters in Berkshire, England, more than 300 years ago. Army veterans told of the remarkable hogs that were larger than other swine and produced hams and bacon of rare quality and flavor. The Berkshire hog became a favorite among the upper class of English farmers, and for many years the Royal Family kept a large Berkshire herd at Windsor Castle.

In the early 1800s, a famous Berkshire boar was named Windsor Castle because he was born to a litter that was farrowed and raised within sight of the royal residence. This boar was imported into the United States in 1841 and caused a stir in the rural press, which reported his mature weight to be 1,000 pounds.

The first Berkshire hogs appeared in the United States by the early 1820s and quickly became popular because of their ability to produce better offspring when crossed with the common stock available at the time. The original Berkshire was reddish or sandy colored and sometimes had spots. Siamese and Chinese hogs were crossed with the basic stock that added the quality of more efficient gains. This cross also helped establish the color pattern we see today: white feet, tip of tail, and snout. In this respect, the color pattern of the Berkshire and Poland China are

very similar. It has been said this is the only outside blood brought into the original breed for the past 180 years, which makes it one of the longest established breeds on record.

In 1875, the American Berkshire Association was created to safeguard the breed's purity. This was the first established swine registry in the United States and only hogs directly imported from established English herds or traced directly back to imported Berkshires are accepted for registration.

The Berkshire is noted for good growth rates, meatiness, and reproductive efficiency, although the litter size is somewhat smaller on average than other breeds. Berkshire meat is darker, has a higher pH, and contains more intermuscular fat or marbling, and better water-holding capacity than other pork. This produces an exceptional flavor and drives the popularity of Berkshire pork among culinary experts in the United States as well as Japan, the largest traditional export market, where it commands a premium price. Berkshire pork is becoming more popular in gourmet cuisine and a growing U.S. ethnic market.

CHESTER WHITE

The Chester White breed originated from Chester County, Pennsylvania. Although first called the Chester County White, the word county was dropped.

The ancestry of the Chester White includes the Yorkshire and Lincolnshire animals of England, along with the Chester breed developed in Jefferson County, New York. The English pigs were imported sometime before 1812 and breeding between these animals occurred until about 1815.

The Berkshire is a British breed imported to the United States in the mid-1800s. It is noted for good growth rates and the exceptional flavor of its meat.

The Chester White breed developed in the United States. It has been a popular breed for more than one hundred years. They are good mothers that fit many different production systems. *Certified Pedigreed Swine Registry*

Sometime prior to 1818, Captain James Jeffries imported a white boar from England to Pennsylvania that was described as either a Bedfordshire or Cumberland pig. This white boar was bred extensively to white Chester, Yorkshire, and Lincolnshire pigs in the county and thus began the development of the Chester White.

As the popularity of the breed spread across the country, four separate Chester White associations developed between 1884 and 1909. F. F. Moore, of Rochester, Indiana, believed having several factions was not conducive to the development and promotion of the breed. In 1911, he worked to combine the various associations and recording offices and the consolidation was completed by 1930, just in time for the Great Depression, when the Chester White Record Association was reincorporated. Today, it is part of Certified Pedigreed Swine, a unified organization of three historic swine associations, the Chester White, Poland China, and Spotted Poland China.

In recent years, more than 60,000 animals have been recorded yearly. This may not be a true representation of the number of Chester Whites used in breeding programs because large numbers are grown in commercial or nonpurebred programs and never recorded. Chester Whites are popular among pork producers because they are good mothers that produce large litters and are typically sound, durable animals.

DUROC

Originally called the Duroc-Jersey, the present day Duroc breed has its origin in the eastern United States. It received a hyphenated name early on because the exact source of the red hogs that went into the foundation of this breed could not be sufficiently identified. It has been suggested that one source for the red or reddish-brown color came from hogs found on the Guinea coast of Africa. Claims have been made that every country to which early slave trading vessels found their way had hogs similar to those found on the Guinea coast.

Another claim states that Columbus brought red hogs on his second voyage to America and that red hogs were also brought by De Soto on his voyages to this country. These hogs would be presumed to originate in Portugal or Spain. It is also possible that reddish-brown hogs in the Berkshire strain may have entered the gene pool of the Duroc breed.

Two distinct strains of hogs helped develop the Duroc-Jersey breed—the Jersey Red of New Jersey and the Duroc of New York. Established in New Jersey prior to 1850, the Jersey Red was developed by Clark Pettit, who suggested the red hogs came from an importation to the state in 1820. Although this date was disputed, they were nevertheless referred to as "red hogs" for many years and gained a reputation for their extreme size, rugged bone, and rapid growth. The name Jersey Reds was first attached to these hogs by the agricultural editor of the *New York Tribune,* who lived in New Jersey, and the name subsequently entered the advertisements of the breeders of this line. Although Jersey Reds could reach an enormous size at maturity, they sometimes lacked quality, were long and rangy, and had coarse hair and bone structure. However, they were favorably accepted by the markets in the area.

The Duroc was originally referred to as the Duroc-Jersey, although the latter part of the name has been dropped in recent decades. The Duroc is noted for fast, efficient gains on less feed, which makes them ideal for many farms. *National Swine Registry*

The Duroc strain was developed by Isaac Frink, who lived in Saratoga County, New York, two hundred years ago. In 1823, he bought some red hogs from Harry Kelsey, who had a famous Thoroughbred stallion named Duroc. When Frink visited the Kelsey farm to see the horse, he took a liking to Kelsey's red pigs and purchased some to take home. Since the pigs had no breed name, Frink called them Durocs in honor of the stallion. It's the only breed of pigs named for an equine. It's recorded that Kelsey told Frink the hogs were imported, but the ancestry, place of origin, or number of generations elapsed from the importations was never established.

The Duroc strain was smaller and more compact than the Jersey Red but had better quality and its offspring had a better rate of gain. In about 1830, William Ensign, also of Saratoga County, bought a pair of red pigs in Connecticut that crossed well with the pigs developed by Frink. Most of the red hogs in Connecticut of that time were said to have been of the Red Berkshire strain that came to Long Island around 1820.

The Duroc-Jersey breeders from Saratoga and Washington counties in New York met in 1877 and developed a scale of points for the breed. In more recent times the Jersey part of the name has been dropped, and the United Duroc Swine Registry is part of the National Swine Registry.

There is a considerable variation in the color of Durocs, which can range from a light golden color, bordering on yellow, to a dark red, similar to mahogany. This color suits many pork producers because a solid color eliminates the concern for proper markings, unlike other breeds.

The typical Duroc has medium length and drooping ears with a slight dish in its face. The breed is recognized and appreciated for its feed efficiency. In many instances it is unequaled by any other breed for its efficient conversion of pound of feed to pounds of red meat, which is one reason Durocs are popular as sires in crossbreeding programs.

HAMPSHIRE

The Hampshire is one of the most easily recognizable breeds. They are black with a white belt running over their shoulders and down their front legs. Hampshires are the third most recorded breed of hogs in the United States and may well be one of the oldest American breeds of hogs in existence today. Historical records seem to indicate that the breed is descended from the "Old English Breed," a black hog with a white belt found in counties bordering southern Scotland and northern England. While criticized for their large size, they were admired for their growth rate, hardy vigor, foraging ability, and carcass qualities.

Hampshire pigs were imported from Hampshire County, England, into the United States between 1825 and 1835. Most of the offspring of these early importations went to Kentucky, where the breed had much of its early development.

The Hampshire developed a following among Ohio butchers who traveled yearly to Kentucky to contract these belted hogs for a premium price. This caused a steady growth in the breed, and in 1893, a small group of Kentucky farmers met to form an association to protect the purity of their breed's distinctive color markings.

The association's first name was the American Thin Rind Association because the members' pigs had skin that was thinner than most other hogs. Known by several names, including Saddleback, Ring Middle, McKay, and McGee, the breed name was changed to the American Hampshire Record Association in 1904. Today it is known as the Hampshire Swine Registry, a part of the National Swine Registry.

Between 1910 and 1920, the Hampshire breed swept across the corn belt, and its popularity continues today because they are productive animals that produce a lean carcass with good muscle quality, large loin eyes, and minimal amounts of back fat. Hampshire females have gained a reputation as being excellent mothers and have the ability to stay in the sow herd a long time. They can adapt to confinement, concrete, or open spaces.

If showing is part of your plans, you should be aware of the rules regarding breed markings for registration. Registered boars and gilts must be black with a white belt encircling the body, including both front legs and feet. They can have white on the nose as long as the white does not break the rim of the nose. When its mouth is closed, the white under the chin cannot exceed what a U.S.-minted quarter will cover. White is allowed on the rear legs as long as it does not extend above the knob of the hock. Registered Hampshires must also have at least six functional udder sections on each side of the underline.

Hampshire pigs classified as off-belts—those not meeting the above requirements—are not eligible for registration with the National Swine Registry. Off-belted animals can still be shown in market hog classes. Further information about these rules can be obtained from the National Swine Registry office.

POLAND CHINA

The Poland China hog, with its six white points, resembles the Berkshire in color but has floppy ears. The Poland China breed had its beginning in the Miami Valley and Butler and Warren counties of Ohio. In 1816, John Wallace, a trustee of the Shaker Society, secured one boar and three sows that were known as Big China hogs from a Philadelphia firm. The boar and two of the sows were white, while the third sow had sandy to black spots. These hogs were reported to be very popular in Maryland, Pennsylvania, and Virginia, and there are claims that the Poland China's origin consisted of six separate breeds.

The swine industry of southwestern Ohio developed quickly between 1816 and 1835 because of the beneficial effects of the Big China hog on farms. They were good feeders, matured early, were prolific, and passed on these characteristics to their offspring. Their size and the fact that they were good travelers made them desirable because they had to be driven to market, which in some cases was one hundred miles.

The Hampshire is one of the most recognizable breeds in the United States and has distinctive markings of black with a white belt over the shoulders and front legs. Hampshires typically produce a lean carcass with good muscle quality, which makes them very popular. *National Swine Registry*

The Poland China resembles the Berkshire with its black color, although it has six white points: the feet, tail, and tip of the nose. The breed leads the United States in pork production. *Certified Pedigreed Swine Registry*

The difference between the Poland China and Spotted Poland China has more to do with hair and skin color than any other physical characteristic. To be eligible for registration in the Poland China Record Association, the animals must possess Poland China Breed characteristics, including being black with six white points: face, feet, and switch (tail hairs). They may have an occasional splash of white on the body but may not possess more than one solid black leg to be accepted as a Poland China. They must have floppy ears, must not have evidence of a belt formation on any part of their body, and they cannot have red or sandy hair or red skin pigment. For registry, the piglets must be ear notched within seven days of birth.

The Poland China is recognized as a big-framed, long-bodied, lean, and muscular individual. This breed leads the United States in pork production. The breed is known for producing good-size litters several times a year, which translates into reproductive efficiency and good mothering abilities.

SPOTTED POLAND CHINA

The Spotted Poland China descended from the Spotted hogs that trace part of their ancestry to the original Poland China, which some believe developed from six separate breeds. One of these six breeds imported into Ohio was called Big China, which was mostly white but had some black spots.

Several breeders in central Indiana brought boars and sows from Ohio into their herds to cross with their own hogs. For many years, they bred these spotted pigs together and developed a separate breed that kept the characteristic color of large black and white spots. At one time in the

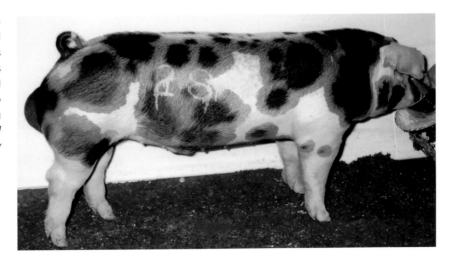

The Spotted Poland China is related to the Poland China but has black spots of varying sizes across its body. It is a popular breed because of its ability to transmit fast-growing traits to offspring. *Certified Pedigreed Swine Registry*

breed's development, two Gloucester Old Spots were imported from England to add a stimulant to the breed in the form of new bloodlines.

Many farmers in that section of the country believed these spotted hogs were superior to those of other breeds and to the Poland China itself. With the breed's growing popularity, the breeders decided to organize and promote Spots as a separate breed, and in 1914 the Record Association of Bainbridge, Indiana, was organized. Today the National Spotted Swine Association is located in Peoria, Illinois, and is part of the Certified Pedigreed Swine group.

The typical Spotted Poland China will have floppy ears and black spots of varying sizes across both sides of its body. Spotted hogs are popular with farmers and commercial hog producers for their ability to transmit their rapid growth traits to their offspring, along with excellent meat qualities. The breed has improved in feed efficiency and the females are known to be docile and easy to handle.

YORKSHIRE

The Yorkshire breed originated in the county of Yorkshire, England, and was classed as belonging to the country's Large White breed in 1884. Yorkshires are thought to have been brought into the United States around 1830 and taken to Ohio.

Following World War I, the Hormel Company tried to promote Yorkshire hogs to farmers in Minnesota and Iowa without much success because the market for lard was vanishing. Farmers breeding Yorkshires had incorporated too much Middle White and Small White breeding into their herds, which resulted in slower growing animals with short, pug noses. The perception of Yorkshires changed with importations of many English Large Whites from the British Isles by some top breeding establishments in the United States, such as the Curtiss Candy Company in Illinois.

With these new bloodlines, the Yorkshires were being transformed into sows with good mothering ability that produced large litters. Modern Yorkshires have more length of body and frame size than their ancestors. It wasn't long before Yorkshires were leading all breeds for rate of gain and feed conversion at test stations. The first herd book was established in Minnesota in 1901, and in 1948, the American Yorkshire Club was reorganized. Today it is part of the National Swine Registry.

As with other breeds that had a transformation in their physical makeup, the Yorkshire breed flourished during the late 1950s and into the 1970s. During the fifteen years between 1957 and 1972, almost 500,000 Yorkshires were registered with the association, compared to a little more than 200,000 in its first sixty-four years. In 1985, the breed had the largest number of recordings of any breed in the United States.

The breed is white, has medium-sized erect ears, and is one of the largest breeds used in commercial production, with a mature boar weighing 600 to 800 pounds. Approximately 80 percent of the world's pork production is based on the Yorkshire breed. Yorkshires have long deep bodies and are known to produce large litters. The sows tend to be good milkers, and the breed produces a carcass with many desirable traits. Yorkshires are commonly selected as one of the maternal lines for commercial pork producers.

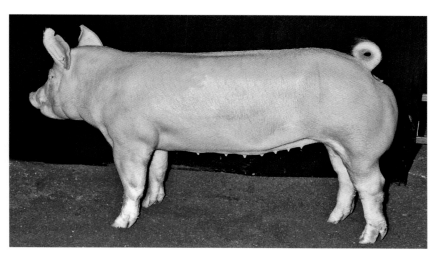

The Yorkshire originated in England and has developed a large following in the United States because of its efficient use of feed and fast rate of gain. *National Swine Registry*

HERITAGE BREEDS

The Livestock Conservancy has placed nine breeds of pigs on its conservation priority list that have either critical or threatened status. Four breeds, the Choctaw, Gloucester Old Spots, Guinea Hog, and Ossabaw Island Hog are not readily available for commercial use and will not be described here. To be included on this list, the breed must satisfy certain genetic and numerical parameters. Of these nine breeds, seven are considered critical, which means there are fewer than 200 annual registrations in the United States and an estimated global population of less than 2,000. Two breeds are considered on the threatened list with fewer than 1,000 annual registrations in the United States and an estimated global population of less than 5,000 animals.

The numerical parameters state that in the United States there must be at least three breeding lines, at least twenty breeding females, at least five breeders in different locations, and that the breed must have an association of breeders.

Although there are several other conditions needing to be met for a breed to be on their priority list, these conditions mean that it is likely there are animals available for anyone who wishes to purchase them to raise. While distance and price may be factors that could limit this acquisition, the fact that they may be available should be looked at as an opportunity to pursue a unique breed for your swine operation.

HEREFORD HOG

The Hereford is a medium-size breed that is unique to the United States. Between 1920 and 1925, a group of Iowa and Nebraska breeders established this breed by combining the Duroc, Chester White, and Poland China bloodlines. By using linebreeding, inbreeding, and a judicious program of crossing the offspring back and forth, this distinctive breed reached about one hundred head of foundation stock by 1934. That year the National Hereford Hog Record was established to promote this new breed, and within ten years the association had attracted 450 members, with most residing in Iowa, Illinois, and Indiana.

The color markings of the Hereford hog attracted the attention of Hereford cattlemen because of the similarity of the markings. Both had a color pattern of intense red with white trim around the edges—the legs, head, and tail. Hereford cattle breeders sponsored the formation of the National Hereford Hog Record because of these unique features.

Although a deep red color is preferred, the shade of red can vary from light to dark. A pig that has a white belt or carries more than the one-third white exclusive of the head and ears is not eligible for registration and cannot be shown in breed shows.

One of the favorable characteristics of the Hereford is that it is very adaptable to a variety of climates and can thrive in outdoor farming practices or in confinement. They are known for their quiet and gentle dispositions, which makes them ideal for

The Hereford has a pattern of intense red with white trim around the edges. This color pattern resembles the Hereford beef breed from which it derived its name. The breed's quiet and gentle dispositions make it ideal for 4-H and FFA projects. *National Hereford Hog Record Association*

The Large Black is a distinctive heritage breed with a loyal following from those who raise them. Although small in total numbers, the Large Black is gaining in popularity because of its mothering ability and its production of lean pork and bacon. *North American Large Black Pig Registry*

young people involved with 4-H or FFA projects, because the breed combines market conformation with an attractive appearance.

Hereford hogs typically have the same general body type or conformation as other breeds. Their medium-size heads have a slight dish and they carry medium-size, floppy ears.

Herefords mature early with many weighing 200 to 250 pounds at five to six months of age. They perform well on pasture but also do well with grain-based feeding programs and reach market weight on less feed than many other breeds. Mature boars may reach 800 pounds and sows about 600 pounds. The sows are excellent mothers, generally produce and wean large litters, and are very attentive to their bright red-and-white piglets.

When the pork industry shifted from the commercial use of purebred hogs to crossbreeding in the 1960s, Hereford hog numbers declined. Today, there are an estimated 3,000 Hereford pigs in the United States, located mainly in the upper Midwest and Plains states. With their inherent characteristics, the Hereford is a natural choice for a variety of small-scale production systems.

LARGE BLACK

The English history of the Large Black combines the black pigs of Devon and Cornwall and the European pigs found in East Anglia, which were developed from descendants of Chinese pigs brought into England in the late 1800s. At the turn of the last century, the Large Black became noticed outside its native region and spread throughout Great Britain.

In the early part of the twentieth century, the Large Black was used for pork production in outdoor systems. Large Blacks have been known to thrive on a diet approaching 65 percent grass, cereal grains, or vegetables. They can adapt and thrive in extreme conditions. Their hardiness and grazing ability make them efficient meat producers. While used mostly for lard production in the past, the Large Black has been bred to produce lean pork and bacon today. Meat from the Large Black is tender, fine-textured, and succulent, making it very desirable for consumers' tastes.

Large Black sows are good mothers and have an excellent milking ability. They typically have good body depth and length, long heads, very large droopy ears, and fine, thick hair. Large Blacks are quite docile and make good 4-H and FFA projects. Although they produce quality pork, Large Blacks do not mature as quickly as most other breeds. However, mature sows can reach 500 to 550 pounds and mature boars can weigh 600 pounds or more.

In 1985 an importation of Large Blacks was made by a group who felt the Large Black, with its ability to be productive in rough conditions, would be an economically attractive breed for farmers in the United States. Because it was not closely related to other breeds already here, it could be used in crossbreeding programs for increased hybrid vigor.

In 1998 another importation of one breeding-age boar and three breeding-age gilts—all unrelated—was made to a farm in New York. Today, there are a small number of herds in the United States with animals descended from the 1985 importation.

The Large Black breed association was formed in 1999, and by 2007, 300 registrations had been made. Currently there are about seven active Large Black breeders in the United States.

MULEFOOT

With fewer than 200 registered Mulefoots in the United States, the Mulefoot Hog is the nation's rarest swine breed. Although its ancestry has unclear origins, it most likely descends from Spanish hogs brought to America in the 1500s.

The Mulefoot is a distinctive American breed that has a solid, noncloven hoof that looks like the hoof of a mule or horse. While the single-gene mutation for Mulefoot can occasionally occur in other breeds, it is a consistent part of the Mulefoot breed in both appearance and behavior because the Mulefoot gene is dominant.

By 1910 there were 235 Mulefoot breeders in twenty-two states. Many of those breeders lived throughout the Corn Belt and in the Mississippi River Valley where they could put their pigs to pasture on islands in the spring and collect them in the fall for slaughter. The Mulefoot is valued for ease of fattening on many kinds of forage, making the dark, moist meat excellent for ham, lard, and home curing.

Mulefoot hogs have mature weights between 250 to 400 pounds for sows and 350 to 600 pounds for boars. They are solid black in color with white points that rarely occur. The sows make good mothers and typically have five or six piglets, but can have up to twelve, and have excellent, calm dispositions.

RED WATTLE

Also on the critical list is the Red Wattle, a large, red hog known for its distinctive flesh wattle attached to each side of its neck. While the wattles have no known function, they are a single-gene characteristic that is always passed to their purebred descendants and usually to their crossbred offspring.

The origin of the Red Wattle is obscure, but some historians believe they were brought to America from New Caledonia, a French island in the South Pacific near Australia, because they were available in French New Orleans in the late 1700s. What exists of the present day herd was developed from animals found in the wooded areas of east Texas. On two occasions, these animals were discovered by farmers who developed a breeding program with the pigs.

H. C. Wengler found his Red Wattle pigs in the early 1970s and reportedly bred two Red Wattle sows to a Duroc boar. He then bred the wattle offspring back to the original sow and developed his herd over several generations. In the early 1980s, Robert Prentice found another wattle herd in east Texas and combined his animals with Wengler's.

Red Wattle pigs brought a premium price during the 1980s. They can produce a lean carcass and show good hybrid vigor when crossed with other breeds. Red Wattle hogs are known for their hardiness, foraging ability, and rapid growth rate. Mature pigs can reach 600 to 800 pounds. The sows are known for having large litters of ten to fifteen piglets and providing good quantities of milk for all of them. Their docile nature and foraging ability makes them a good choice for use in outdoor or pasture-based swine operations, and they can adapt to a wide range of climates.

The number of Red Wattle pigs in the U.S. is relatively small and they may be difficult to find.

However, the Red Wattle breed has a registry association even though the breeding sow herd in the United States is about 300 and there are about six to ten active Red Wattle breeders.

TAMWORTH

The Tamworth breed originated in central England and takes its name from Tamworth, Staffordshire, where Robert Peel developed his herd from pigs he imported from Ireland. Known as the Irish Grazer, the long and lean Tamworth is perhaps the most direct descendant of the native pig population of northern Europe. Being an outdoor pig, they were expected to find their own food in the forests and subsisted on acorns and whatever else was available.

The breed was brought to America in 1882 by Thomas Bennett, of Illinois. During the next five years, other Tamworths were imported to Canada and brought to the United States as well. The American Tamworth Swine Record Association was formed in 1887, and by 1920 the breed had become established across most of the United States and Canada.

The Tamworth is a long, lean animal considered a better bacon than fat producer. During the eras when lard was highly prized, it didn't gain in popularity and its numbers considerably dropped. However, this breed is favored by bacon curers because of its length and depth of sides, light waste, and the large proportion of lean red meat to fat.

The Tamworth originated in England and is not directly related to many of the other breeds, which makes them ideal for crossbreeding. The Tamworth produces highly desirable bacon products from low-input production programs. *Tamworth Swine Association*

23

Because they can be considered a bacon breed, Tamworths thrive on low-energy foods and have a slower growth rate. It is not as competitive in a confinement program and doesn't seem to do well when placed with aggressive breeds. However, this makes them very useful in small-scale pig programs that use outdoor feeding systems. They can be used for salvaging crops or turned out on a pasture after it has been grazed by cattle. The Tamworths' long heads and snouts make them good foragers, and their long legs allow them to walk long distances. They are a good fit for pasturing and low-input operations.

The Tamworths have a red, almost ginger, color to their hair coat. Some are a dark red, which makes them adaptable to many different climates, and they are tolerant of extreme changes in temperature. Their ears are erect and medium sized. Mature pigs are typically medium in size but can reach weights of between 500 and 600 pounds for sows and boars. Sows make good mothers, typically have large litters, and produce sufficient quantities of milk for their piglets. Because their meat is lean, the Tamworth is a good choice for niche meat markets of pasture-raised, lean meat of excellent quality.

OTHER BREEDS

Several other swine breeds exist in the United States but have few commercial opportunities for varying reasons. For example, the Choctaw breed has long been associated with the Choctaw Nation of southeast Oklahoma. Guinea Hogs have a population of fewer than 200 and generally have small bodies that contain a high percentage of body fat, which makes them less desirable by processors and consumers.

The Ossabaw Island Hog, named for the island off the coast of Georgia on which they developed, have a form of low-grade, noninsulin–dependent diabetes. This condition resulted from the hogs adapting to the food cycle on the island that provided little to eat during the spring and caused them to develop a unique biochemical system of fat metabolism that enabled them to store a larger proportion of fat than any other hog. While making excellent medical research animals, it is impossible to import them directly from Ossabaw Island due to quarantine restrictions, although small breeding groups—descendants of animals brought from the island during the 1970s—can be found on the mainland.

Crossbred pigs result when parents from two different breeds mate. In many cases they can be easily identified by the mixture of colors on their hair coat. Crossbred pigs cannot be registered with breed associations.

NATIONAL SWINE REGISTRY

The National Swine Registry (NSR) is a cooperative group that was formed in 1994 as a result of the consolidation of the American Yorkshire Club, the Hampshire Swine Registry, and the United Duroc Swine Registry. In 1998, the American Landrace Association joined the group and these four breed organizations, which represent 87 percent of the total purebred hog population in the United States, are located in one central office in West Lafayette, Indiana. This group provides educational materials, such as swine-judging videos, free genetic consultations and evaluations of herds, and litter registrations and marketing assistance.

CERTIFIED PEDIGREED SWINE

Formed in 1997, Certified Pedigreed Swine (CPS) is a unified organization of three historic swine associations: Chester White, Poland China, and Spotted Poland China. This group shares a common interest in promotion, breed purity, performance programs, and genetic advancement. Together the group is able to offer affordable services to members that they could not offer separately. The organization maintains ancestral files, performance records, and membership lists, and conducts breed shows and junior activities across the country.

BREEDS OF PIGS

CERTIFICATE OF REGISTRY
HEREFORD
An Iowa Corporation Not For Profit.
NATIONAL HEREFORD HOG RECORD ASSN.

THIS IS TO CERTIFY that the pedigree of the

SEX.......... NAME........................ NO.............

Farrowed .... Number in Litter........ Number Raised........ Boars........... Sows..........

Description .... Ear Marks:

Bred by ....

Owner [when farrowed] ....

Sold to ....

SIRE:................................. By............

Bred by......................... Out of.............

DAM:............................... By............

Bred by......................... Out of.............

has been accepted for entry in the NATIONAL HEREFORD HOG RECORD under the official number indicated above. This entry is based upon the certified statement of the owner of the animal when farrowed, supported by the certified statement of the owner of the dam at the date she was bred to produce this animal, in case such first owner and breeder were not the same person.
GIVEN under the hand of the Secretary of the National Hereford Hog Record Association and the seal of the corporation at Flandreau, South Dakota.

Date ....   SECY.

Transfer of ownership to be indicated on the back of this certificate.

Registry associations provide standard forms that are completed with the required identification information and returned to the association's office for processing. In return you receive an official certificate identifying this particular animal as registered with the association. This form of identification helps to construct a pedigree or ancestry chart of a pig.

No. 19—   Fall Boar
        Farrowed August 28.   Litter of 8.
  Bred and consigned by Howard Hasheider, Sauk City, Wis.

Smooth Rex 635959 .... { Hillside Spotlight 624969
                                 { Supreme Sara 771624

Heiderland's Penny 997686 .... { The Foreman 617953
                                { Heiderland's Leona 952316

This is a typical pedigree that can be seen at a registered pig sale. It shows the pig's ancestry, along with other information, including the farrowing date (when it was born) and how many pigs were in that particular litter. The registration numbers behind the names verify that those ancestors are registered with that particular association.

Raising critical or threatened status breeds, such as these Red Wattles, may enhance your program as you become part of the effort to rescue them from extinction. Even very small associations have registry programs to help identify pigs of the breed. However, producers and growers who obtain pigs from breeds with few numbers should be encouraged to continue developing them as pure lines.

OTHER BREED ASSOCIATIONS

Registry associations exist for all the heritage breeds mentioned. Their contact information can be found in Appendix 2 (page 172). If you are interested in raising one of the breeds listed as critical or threatened, you should contact the association. Breeders of a particular swine breed tend to be passionate about the pigs they raise and their enthusiasm can be infectious. Helping save a breed from extinction is a worthy cause that can significantly add to your family's satisfaction in raising pigs.

If you choose to raise a critical or threatened breed, you are encouraged to maintain the registrations of their offspring with the specific association. For many of the small breeds this is the only way to increase their foothold in the swine population. As more people choose to raise these breeds there is a greater chance to pull them back from the brink of extinction.

The Yorkshire is easily identified by its erect ears and white body. The breed's popularity has resulted because of the many desirable traits the pigs pass on to their offspring. *Tom Rake*

The Chester White breed records a large number of registrations each year because of its popularity with growers. Many Chester Whites are used in commercial and purebred programs, as well as in crossbreeding. *Tom Rake*

Sows that can produce a large number of piglets each litter are the most profitable. A good nutrition program, close observation at farrowing, and a clean, sanitary pen will help ensure a high piglet survival rate. *Tom Rake*

The purebred Hampshire has an attractive and distinct color pattern. Hampshire sows make excellent mothers and are recognized for their ability to stay in the herd for many years. *Tom Rake*

A crossbred boar can be identified by the multiple color markings across its body. Crossbreeding programs make use of two or more different breeds. One of the advantages of crossbreeding is the increased hybrid vigor that typically gives the offspring a hardy start after they are born. *Tom Rake*

Regardless of the breed or breeds you choose, the goal for any pig-raising program is to achieve high rates of pregnancy, piglet survival, and growth. One of your most important decisions will be to choose a breed that fits your farm and goals. *Josh Wendland*

CHAPTER 2
GETTING STARTED

To get started in raising pigs you'll need land, buildings, pigs, and equipment. Purchasing a farm generally involves purchasing a business as well, because there are financial considerations whether you work the land, rent it to another party, or leave it to lay fallow. How you handle these options may have much to do with your financial situation, inclinations toward farming, and level of involvement in the farm. If you already live on a farm but do not have any animals, you may decide that raising pigs is a viable option.

Farming can provide a rewarding financial and emotional experience to you and your family. It is a unique way of life.

When purchasing or renting a farm, factors such as location and size of the farm, soil type, house or dwelling, buildings available, and a number of other intrinsic factors including schools, social outlets, and a sense of community may be important to you and your family. Planning, research, and obtaining good advice will help you avoid unpleasant surprises when purchasing a farm. Advice for purchasing or renting an available farm can come from an agriculture lending group or bank, a county agricultural extension office, or private professional services that specialize in farm purchases and setting up farming enterprises.

You can do much of the initial research on your own by contacting real estate agents about the availability of farms for sale or rent or by visiting properties on your own in locales where you may want to live.

LOCATION AND SOCIAL CONSIDERATIONS

In many cases your farm will also be your home. The property's location and the services available may be important factors when deciding where to buy. Living in a rural area is not the end of the world; however, there are some significant geographic differences between rural settings and urban ones. Living on a farm does not necessarily exclude you or your family from the conveniences or services available in urban areas. There just happens to be a greater distance to access them.

When purchasing a farm, assess whether the house or dwelling meets your family requirements both now and in the future. If you have a young family, it may be important to be close to schools, doctors, or transportation systems. The availability of professional veterinary services may be important for your animals. If community activities are important, you can visit the area's chamber of commerce, which provides information about local events during the year. You may also want to look at the opportunity for alternative or off-farm income or employment.

PHYSICAL CONSIDERATIONS

The goals you set will determine the size of the farm you need to have. You do not need a large-scale farm to raise pigs. Pigs fit many types of small-scale enterprises.

Soil type and fertility are important considerations and may be tied to property values. Soil type also influences crops raised and their durability during a drought or extended dry spell or extremely wet conditions. Heavy soils sustain crops better in dry conditions than lighter, sandier soils.

The quality of buildings and improvements may be a determining factor in your purchase. Extensive building renovations require finances that could be directed toward farm operating expenses. Yet the need for improvements may lower the purchase price and be an attractive option.

Whatever farm you purchase, it is necessary to fully understand its boundaries. Walking the fences provides you with an idea of how much land there is, as well as information about the condition of the fences, buildings, and the soil and other aspects of the property, such as suitability to pasture-raise pigs.

Prior to purchase, be sure to check for the presence of any contaminants or residues that could affect the health of your family or livestock. Taking a test of the farm water well is a good idea. Underground fuel storage on farms has been banned in most states. Old storage tanks may still be present and need to be dug up and removed. Be sure to address this issue prior to signing any purchase agreement.

Living close to the land can open new possibilities for a business developed from raising pigs. Good planning and research will help you avoid unpleasant surprises.

Your buildings do not need to be elaborate in order to raise pigs, but sound structures are essential for safety concerns relating to your family and animals. Remodeled buildings may suit your purposes and still provide a comfortable place for working.

BUYING YOUR FIRST PIGS

There are several ways to purchase pigs, and it is wise to have a plan before you start. No matter which livestock species you wish to raise, buying animals always contains a certain amount of risk. You can lower this risk by considering several factors in purchasing animals, including their overall physical condition, health, and mobility, and knowing their source. These may not be the only criteria you use in your considerations, but they will provide a foundation for selection.

Physical condition refers to the weight the animal carries on its body. In the case of feeder pigs this will be very small, while older pigs should have weight ranges consistent with their age. Overly fat pigs will cost more, especially if you later put them in pasture and they lose much of this weight. It is best to buy lean pigs and have them gain weight based on your feeding program.

Overall health of the pig is more important than physical condition, although the two can be closely linked. A healthy pig typically is an aggressive eater, alert, and adventuresome. The quickest way to determine the health of any pig is simply to look at it. Does the animal appear to be alert? Does it have clear, dry eyes? Is it sneezing or coughing?

Thin or emaciated pigs should be avoided entirely, no matter what the price, because this may indicate serious health problems. Avoid pigs with a dull appearance, discharges from their eyes or nose,

breathing abnormalities, listlessness, or anything else that strikes you as abnormal.

Another health consideration is mobility. Look at how they move around. A pig should have the ability to move about freely with no leg, joint, or feet problems. Avoid pigs that limp or have swollen joints, long toes, or other physical impairments. Pigs exhibiting any of these conditions will not last long on any farm. If you are not sure of your own expertise at identifying problems, hire a veterinarian or someone with experience to go along to look at the pigs prior to purchase. This will be money well spent.

Knowing the source of your pigs may alleviate many worries about their health. If you purchase pigs at a farm, take a look around the farmstead. If it is well-kept and clean, it is likely the farmer pays attention to the details of his pigs, as well. Observe the attitude of the person selling the pigs as this can often provide clues as to their treatment and care. Having the satisfaction of buying in a pleasant surrounding from a caring farmer can ease your concerns about the pigs you buy.

Depending on which phase of swine production you wish to enter, you may want to consider buying pregnant sows or feeder pigs to begin. This provides you with an opportunity to start at the beginning of one process and quickly work your way toward the end product. It also eliminates the need for securing a boar right away.

If you have little or no experience with raising pigs, you may want to consider starting with a small number, whether they are pregnant sows or feeder pigs. This minimizes your initial investment, requires less labor, and allows you to familiarize yourself with the pig-raising process. As you gain experience, confidence, and expertise, it will be easier to plan for more pigs.

START-UP ECONOMICS

Using certain criteria and assumptions, you can calculate a potential starting cost. For example, you may buy feeder pigs and raise them to market weight. At this writing, feeder pigs are selling at $35 for 40- to 50-pound pigs, making an initial investment—for say twenty pigs—of $700. At current prices, breeding sows can be purchased for about $150 to $200 each, boars for $200 to $250 each, and gilts for $150 to $200 each. These prices reflect strictly commercial pigs and not registered animals that may cost more because of their value as breeding stock.

Feed costs will be the largest expense during the growing period between purchase and market. You may choose to raise your own crops or purchase feed. Typically hogs require between 3 to 5 pounds of feed per pound gain through the three phases of production—nursery, grower, and finisher. You can calculate the estimated total feed costs by multiplying 800 pounds times the cost of the feed times the number of pigs involved.

These figures will need to be adjusted to fit your situation and current crop market conditions. Better feed efficiencies will result in lower feed costs, and pasture-raised pigs may take longer to reach market weight because of a slower growth rate than hogs raised in confinement.

Unless you plan to absorb all the expenses for the pleasure of raising pigs, you may want to develop a value-added marketing program to offset the production costs. This could involve organic production, which commands a higher market price, or pasture-raising, which can lower production costs, or a combination of both.

The best way to maintain a robust herd is to purchase healthy, physically sound pigs at the beginning. Healthy animals are easier to raise and grow better by using feed more efficiently. Healthy pigs can thrive in all kinds of weather and climates with proper management.

One advantage of buying pigs privately (from another farmer or pig grower) is that you reduce the risk of exposure to many different farms, any of which may have sick animals.

WHERE TO BUY PIGS

As with other livestock, pigs can be purchased through two venues: public or private. They can be purchased privately from another farmer or specialized pig grower. Public sales include feeder pig sales, auction barns, or on the Internet.

Each venue has its advantages and disadvantages. A public auction is where any member of the public can bid on animals to purchase. In these auctions, such as feeder pig sales, you will pay a price that is the highest bid and you will be expected to present a good check after the sale. One advantage of buying at a public auction is that there are established conditions under which pigs are sold and sometimes guarantees are stated at the beginning of the sale regarding the animal. These guarantees could include details that may be important to you.

Another advantage of an auction market is that the price for those animals is determined by other bidders. This can provide a reasonable assessment

of the worth of those animals by other, perhaps more experienced, pig growers and can affirm your judgment of the animals. But in some cases, the price may be more than you want to pay and you go home empty-handed. Feeder pigs generally sell in groups at a public auction. This may eliminate the need for you to attend several sales to get the number of animals you want to start with.

The final bid at an auction is the price you pay and there are usually no negotiations at the end of the bidding. Unless there is a major problem discovered after the purchase, you generally can't return the animals. It doesn't mean, however, that you can't discuss any problem with the seller at a later time to arrive at an agreeable solution. Most established pork producers are willing to set things straight if problems arise, especially if they know that you are a new customer. It is in their long-term interest to have satisfied customers.

This is one of the advantages of purchasing animals privately. You may pay a price that is determined by the owner, but you also may be able to get all your animals at one place and reduce the risk of exposure to other pigs from different farms. When buying privately it may be possible to negotiate a better price if a large number of animals are involved in the purchase. An established producer's reputation is important, and most will try to accommodate a buyer's concern. You may also be able to get replacements for those animals that develop problems.

BUYING A BOAR

Buying a boar is a different situation from buying feeder pigs or females for breeding purposes. The feed efficiency of your pigs will be determined in large part by the genetic inheritance from their sire and dam, and the fastest way to bring in new genetics without replacing the entire female population is using a superior boar. Whether you buy a boar for breeding purposes or raise one of the male pigs from a litter may depend on your situation. Many popular crossbreeding programs use a rotational plan that produces replacement

females so only boars need to be brought into the herd. This also keeps the herd exposure to outside health concerns to a minimum.

Whether you are producing pigs for the general market or for a niche market, growth and carcass traits will still be a consideration. Try to select a boar from a line of pigs with a positive genetic influence. This may not be possible in some heritage breeds where the genetic pool is limited and genetic advances for carcass and growth traits may be slower.

Finding a supplier of boars may become easier once you have determined your needs. When looking at a boar, you should inquire about the health status of the herd, vaccination schedules, whether there have been any disease outbreaks, if they routinely feed antibiotics, and if so, why. The owner should be willing to provide you with information about health testing that may have been done by a qualified veterinarian, or ask for the name of the herd veterinarian with whom you can discuss any concerns. If they don't provide you with this information, you may be left to your own judgment on the health of the herd.

If something goes wrong with pigs you bought at a public auction, you generally can't return the animals. Nevertheless, you should contact the seller and discuss the problem: most producers will set things right, as they don't want to develop a bad reputation.

When selecting a boar, you should reach an agreement and secure a guarantee about his fertility and virility. It is important that the boar you purchase be able to impregnate your sows and gitts or you will lose time and money in your breeding program. If a boar fails to settle the females or fails to show interest, a replacement will quickly be needed to maintain your farrowing schedule. Avoid purchasing boars from an owner when an agreement about replacement or a fertility guarantee is not available.

When privately purchasing a boar, you should also consider what the sale price includes. Does it include any breed registration and transfer papers? If so, who will pay for those? Is transport and delivery included or are there extra charges for these services? Will the owner provide you with health certificates assuring you that the herd is either accredited free of brucellosis and pseudorabies or that appropriate blood tests will be done thirty days prior to change of ownership? Securing the appropriate health certificates will ensure that you are bringing a healthy boar onto your farm.

After bringing a new boar onto your farm, you should isolate it from the herd for thirty days and check daily for any signs of illness. During this isolation period, you can pen the sows and gilts to be bred near the boar to observe its sexual interest and libido. At this point it may be possible to check for any physical deformities of the boar's reproductive organs. It is advisable to start any new boar out slowly and not have it breed several sows or gilts in a short time period. Having a controlled environment at first will make the experience for the boar more desirable and may lead to fewer breeding problems down the road.

TRANSPORTATION COSTS

One overlooked cost of buying pigs privately or at public auction is the issue of transporting them to your farm. If you do not have your own transport, alternatives, such as hiring a truck, will need to be arranged. As the buyer, you will be expected to promptly remove the animals from a public auction.

Adjustable loading chutes will allow for easy movement of pigs onto trailers or larger trucks. They can be constructed of wood or metal and can either be permanent or portable structures.

A carrier for small- to medium-sized pigs can easily be built. These can be useful in moving pigs from one place to another without them escaping. If used for transport on a highway, they offer protection and safety during the trip.

If you can hire a trucker to haul them to your farm, make certain you understand the terms for hauling and get a firm estimate. Usually the cost is based on each mile of transport, and depending on the individual truck driver, it may be quoted per mile or by the load. If hiring transport, you should insist on the trailer being thoroughly sanitized before loading your pigs at the point of purchase. If buying privately it may be possible to have the seller transport them to your farm. This may be part of the negotiated total purchase price.

ANIMAL IDENTIFICATION

Identification of animals you have purchased may be important so you can be assured the pigs arriving (if you don't transport them yourself) on your farm are the ones you purchased. Some pigs may have ear tags or ear notches. Take note of them as this can verify that they are the ones you purchased.

Whatever form of animal identification you use will help with your record keeping system. This will be useful in any breeding program you develop to identify good maternal lines, or to avoid the negative effects of inbreeding or linebreeding. This will be much easier if you can positively identify each of your animals over several generations.

HEALTH REGULATIONS AND CONCERNS

Before the pigs arrive on your farm it is important to know their health status. At a public auction some pigs may be sold with health certificates issued by the seller's veterinarian. Such a certificate is important, as it assures the buyer that the pig is healthy and as free of any disease as reasonably possible. If the pigs are purchased privately, try to secure health certificates from the owner prior to leaving the farm. Be sure to fully understand who pays the cost of having the health tests done while they are still at the seller's farm.

Before purchasing any feeder pigs, boars, sows, or gilts, check with your local veterinarian about any health regulations that may apply. Most states require that pigs moved from one farm to another be found negative for brucellosis and pseudorabies within thirty days before change of ownership.

Plan to isolate your boar or any pigs you purchase for thirty days once they are brought onto your farm. During this time you can conduct another blood test for brucellosis and pseudorabies. Some states require a second test for breeding stock while in isolation, so check with your local veterinarian about the regulations in your area.

FARMING PRACTICES—ORGANIC, SUSTAINABLE, OR CONVENTIONAL?

Organic, sustainable, biological, and conventional systems comprise most animal-production systems. The organic, biological, and sustainable systems are gaining popularity with farmers for several reasons. The most prevalent are increased markets for products from these systems and increased value those products bring to the marketplace. There are significant differences between organic and sustainable farming and conventional farming practices. By understanding how these systems may best be utilized on your farm, you can decide which one is best to achieve your goals, increase your income, or support your philosophy.

CONSUMER TASTES

Consumers' preference for inexpensive food drove conventional farming practices for many years. The use of intensive planting and harvesting of crops—aided by crop chemicals and commercial fertilizers—to increase production and generate higher profits has significantly altered the farming industry in the last sixty years.

Consumers are increasingly interested in where their food comes from and how and under what conditions it is produced. Organic, biological, and sustainable production systems are rapidly gaining popularity with farmers and consumers alike. Shifts in consumer tastes and a higher demand for foods produced locally and humanely are making these production systems a viable alternative to intensive pork-production practices.

This consumer shift, which is cultivating the image of healthier foods, has brought more money to organic and sustainable markets. It has allowed farmers to utilize practices that may be more in tune with their personal ethics and nature's harmony. Organic and sustainable farming can be a good fit for small-scale pork production.

No matter which production system you decide to use, you have the responsibility to raise your pigs as humanely as possible and provide them with sufficient food and shelter.

Organic farming is an attractive alternative to conventional farming because it encourages and sustains the natural processes of the soil without the use of synthetic chemicals. Organic, sustainable, and biological farming methods can be successfully adopted in most areas.

ORGANIC AND BIOLOGICAL FARMING

Organic and biological farming are holistic, ecologically balanced approaches to farming that support, encourage, and sustain the natural processes of the soil and animals. They are parallel systems with a perspective that reaches beyond the present and requires a long-term commitment to the program.

Organic and biological farming emphasize management practices over volume practices and consider regional conditions that require systems to be adapted locally. They support principles, such as environmental stewardship. In many ways, both have been a reaction to large-scale, chemically based farm practices that have become the norm in food production since World War II.

Farming organically or biologically are methods where the whole ecosystem of the farm is incorporated into the production of animals or crops. They both rely primarily on contributions from on-farm resources, such as animal manures, composts, and green manures for soil fertility, while eliminating external or synthetic additives.

Organic and biological farming practices are flexible enough so that if sufficient quantities of materials cannot be produced on the farm, then off-farm nutrients, such as natural fertilizers, mineral powders, fortified composts, and plant meals from approved sources can be applied without risking the farm's certification.

Organic and biological farming promote the health of the soil by encouraging a diversity of microbes and bacterial activity. This, in turn, enhances the growth of crops, such as corn, soybeans, grasses, and other crops that may enhance swine health. There is a beautiful symmetry to this type of farming where feeding the soil feeds the plant that feeds the pigs that feed you.

39

Pigs Are What They Eat

The old saying you are what you eat also holds true for livestock. Pigs fed grains and plant proteins grown without synthetic chemical assistance will not ingest these chemicals into their systems. By eliminating synthetic chemical usage on soils and crops, organic and biological farming systems are gaining in popularity because of the perception that these systems produce a more natural and healthy food product. Additionally, one often overlooked benefit from organic or biological farming is the increase in crop pollination assistance from bees and other insects.

Grass-based farming practices protect the soil from erosion. Crop rotations help build up the soil by using a diversity of crops. Rotating what is planted on a field from year to year will help maintain good soil fertility.

Some farmers are attempting to reverse the industrialization and standardization trends that many critics point to as transforming farms into factories. Farmers involved in organic and biological farming support the belief that, as living beings, livestock cannot be molded into an assembly line production model and sustain it for a long period.

Large cropping and confinement farming systems are like assembly lines with raw materials coming in and finished products going out. Organic and biological farming proponents seek to break this cycle by linking their philosophy of life as a physical characteristic of the foods and goods they produce.

Organic Certification Processes

The term organic is defined by law in the United States, and the commercial use of organic terminology is regulated by the government. Certification is required for producing an organic product that is packaged and labeled as such for

Learning centers across the country differ from university-based research programs because they are privately funded and offer a variety of alternative views and perspectives. Open discussions and questions are encouraged, and field days allow visitors to study the results, determine their own conclusions, and consider how these practices may be applied to their own situation.

Producers who venture into organic farming practices support the belief that, as living beings, livestock cannot be molded into an assembly line production model and need to be allowed a more traditional setting for raising their offspring.

sale. While the process to become certified can seem extensive, the result is that products produced under certification are authenticated.

Standards for organic certification are set by the government or by various organizations in which farmers become members. Lists of these organizations are available from most county agriculture extension offices or from many states' departments of agriculture. These organizations can provide information relating to organic certification requirements and markets.

The biological farming process does not require certification to produce products. These products cannot be labeled as organic without submitting to a certification process. They are parallel programs in many ways, but only the organic label requires a certification process.

If you choose to farm organically or biologically, you need to study your soil and have it analyzed to determine which nutrients need to be added to balance its components for optimum plant growth and nutrition. You will be feeding the soil so that it can feed the plants. Under organic rules, in order to maintain your certification, grains fed to your pigs can only come from organic sources. It is important to understand that it is impossible to become certified organic for pork production if crops on your farm are not produced under organic protocols.

If your farm has used conventional practices with synthetic chemicals, there is a three-year transition period before organic certification can be obtained. During this time no synthetic chemicals of any kind can be used on the land or on your pigs, antibiotics cannot be used to treat the pigs, and crops identified as genetically modified organisms (GMOs) cannot be planted.

The exclusion of insect and pest control products, as well as antibiotic use, can make raising pigs seem more of a challenge, but hogs are successfully raised this way in many different parts of the country because farmers are finding new markets for their pork products.

SUSTAINABLE FARMING

Sustainable farming is a goal rather than a specific production system. Although they are often thought to be the same thing, sustainable and organic farming are not necessarily synonymous. The goal of sustainable farming is to approach a balance between what is taken out of the soil and what is returned to it without relying on outside products. The idea of sustainable farming starts with the individual farm and spreads to communities affected by this farming system.

Growing and harvesting crops removes nutrients from the soil, and without replenishment the land becomes less fertile. One of the major keys to success in sustainable farming is soil management. A healthy soil will produce a healthy crop that has optimum vigor and is less susceptible to insect and pest damage.

If you are buying a farm or if you are already farming, it is likely you will need a transition period to become sustainable. This transition process generally requires a series of small, realistic steps. Because of some costs involved, your family economics and personal goals may influence how fast or far you will go in this transition.

Raising pigs using sustainable production principles will include consideration of how you integrate your crop and animal systems. The key components are profitability, management of your pigs, and stewardship of the natural resources, such as water and soil. Animal health is an important part of a sustainable farm, because the health of your pigs will greatly affect reproductive performance and weight gains of growing pigs. Pigs that are not healthy waste feed and require additional labor. While not impossible, raising pigs in a sustainable production system is extremely difficult because of the need for grain-assisted rations. Raising grains typically requires equipment using fossil fuels, which are nonrenewable resources.

Sustainable farming includes attention to the stewardship of natural and human resources. Stewardship of people includes consideration of social responsibilities, such as working and living conditions of farm laborers, the needs of rural communities, and consumer health and safety both now and in the future.

Sustainable agriculture requires a commitment to changing public policies and social values, as well as preserving natural resources. This may

Organic, sustainable, and biological farming promote the health of the soil by encouraging a diversity of microbes and biological activity. Grasses and other plants thrive in such an environment, and multiple species sown in a field can yield striking results.

Organic food production systems can reach many niche markets. Whether you market your pork products from home or at a store may depend on your goals, finances, business plan, amount of products available, or other factors. Developing a market takes a commitment but can have many rewards.

include becoming involved with issues such as food and agricultural policy, land usage, labor conditions, and the development and resurgence of rural communities.

Developing a strong consumer market for your products can help sustain your farm. Becoming part of a coalition can be useful in promoting your products and educating the public in general. By educating more of the public, you will likely increase your market and therefore continue the cycle you started.

Practicing and implementing sustainable farming can be a challenging but rewarding option. For those who believe in preserving the land and water for our use and for future generations, this practice can have benefits far beyond any monetary compensation.

CONVENTIONAL FARMING

Many conventional farming practices are geared for maximum production by using large amounts of outside resources to produce their products. Whether it is corn, soybeans, alfalfa, wheat, or any number of other crops, intensive farming practices have become the norm for many farms. These farms typically use synthetic chemical products to promote rapid growth of pigs and to control insects

and weeds. Highly specialized machinery does most of the work, and the operator's feet may seldom touch the ground.

Conventional farming differs dramatically from organic, biological, and sustainable farming in that there is a heavy reliance on nonrenewable resources, such as synthetic fertilizers, gas, and diesel fuel. There is also the concern that some practices, including excessive tilling, can lead to soil erosion, which may cause long-term damage to the soil.

Although farming intensification may not seem acceptable to certain groups, there are many large-scale producers who are responsible stewards and manage their production ethically and with sound environmental practices.

OTHER CONSIDERATIONS

One of the most attractive aspects about different farming systems is that you can stop one system and switch to another system. It is easier, however, to switch to a conventional farming system from one that is organic, biological, or sustainable rather than the other way around. The option to switch farming systems may be important if a change is required in your production practices to adjust to new circumstances or markets sometime in the future.

PORK-PRODUCTION SYSTEMS—FARROW-TO-WEAN, WEAN-TO-FINISH, AND FARROW-TO-FINISH

There are four distinct phases in pork production: farrowing, weaning, growing, and finishing. Farrowing involves the birth of a litter of piglets until they are weaned at between three to four weeks of age. Weaning, growing, and finishing are the following steps and involve the growth and development of the pig until it reaches a predetermined weight that corresponds to a certain carcass composition.

These phases involve different feeding and housing systems that allow you to specialize in one or more stages of production, depending on the size of your farm, facilities available, and personal inclinations. There are three options you can consider for raising pigs: farrow-to-wean, farrow-to-finish, and finishing, as well as several management production systems, including pasture-grazing, full-feed, and confinement.

The time requirements for each of these systems vary and may be a factor in which one you choose to use. Once a gilt or sow is pregnant, she will farrow (give birth) in approximately sixteen weeks. The piglets are nursed for about three to four weeks before they are weaned or separated from the sow. From that point on, you will have to provide the bulk of the nutrition for your pigs to grow at acceptable rates.

FARROW-TO-WEAN

The farrow-to-wean production system involves breeding and farrowing sows, raising the piglets until they reach a certain age or weight (usually about forty pounds), and selling them to another producer or finisher. This program requires that you keep a herd of breeding sows to replenish your supply of piglets. This can be one of the advantages to your program, because it allows you to keep the sows without having to raise all the young pigs to market weight. A second advantage of this system is the decreased need for facilities and operating

A farrow-to-wean production program allows you to maintain a breeding sow herd without raising the offspring to market weight. This pork-production system is attractive if you don't have sufficient facilities or crops to raise a large number of pigs.

capital, feed, or manure. This lowers the cost of production, because you do not need to provide large amounts of feed to the young as they will be marketed. The feed you save by not finishing out these young pigs can be fed to a larger number of sows in the breeding herd.

This option is attractive if you do not have enough facilities for raising pigs past a certain age or weight or if you do not have enough feed to finish out pigs born on your farm. The number of sows used in this option can vary according to the size of your farm, facilities for farrowing, and your labor supply.

One disadvantage of this system is that producers with small herds are generally at risk during a volatile feeder pig market. With few options available for marketing feeder pigs, you may be caught in high or low cycles depending on many factors beyond your control. Observing pig markets and market trends may provide information to allow you to increase your production of litters if a shortage is foreseen. Similarly, you may wish to decrease production if the market appears to become saturated at certain times of the year.

During this phase, particularly if you plan to continue to raise the piglets, they are offered a creep-feed ration to supplement their mother's milk. This is a high-protein diet that allows them to increase their dependence on feed and lessen their dependence on their mother.

Much of the profitability of a farrow-to-wean system depends on the number of piglets raised and subsequently weaned for sale. There are several factors that will influence this, including the number of piglets farrowed, survival rate of the piglets either at birth or through the first three to four weeks, and quality of the piglets from a genetic standpoint. Paying close attention at farrowing times will ensure optimum survival rates, and providing good management after farrowing will diminish piglet loss and increase the number for sale. The more piglets you have to sell at weaning, the more profits you will receive.

FARROW-TO-FINISH

A farrow-to-finish operation is similar to the farrow-to-wean program with the difference being that you continue to raise the young piglets through to a predetermined market weight, typically at about 240 pounds. This entire production period takes between ten to eleven months, with four months for breeding and gestation, plus six to seven months to raise the litter to market weight.

Of these three systems, farrow-to-finish operations have the most flexibility and typically offer the best potential for profit over a long market period. They can provide a cushion during low market price periods, as well as provide steady income during high price periods. Farrow-to-finish operations generally require more facilities, depending on the size of the breeding sow herd, more feed because of the increased number of pigs that must be fed over a longer period of time, more labor, and a longer commitment on your part.

It is generally best to separate areas for farrowing and for raising pigs to market weight. You may use one barn or shed for farrowing and a different shed or pasture for raising pigs to market weight. One requirement for this option is to maintain a group of sows, which are bred naturally or by artificial insemination. The number of sows to keep will depend upon such things as the size of your farm, capital investment, and labor supply.

This option requires more management because of the different sizes of animals involved. Using separate areas for sows and the pigs being raised for market promotes less competition from larger animals, allows smaller ones easier access to feed, and eliminates dominance at feeding and watering times.

WEANING-TO-FINISH

This option includes raising pigs from weaning through finishing or market weight. Most feeder-to-finish producers buy feeder pigs weighing thirty to forty pounds and feed them to market weight. Facilities such as barns, open sheds, shelters, or other simple structures where the pigs can get out of the elements are sufficient for this phase of swine production.

Some advantages are the low overhead costs, low labor requirements, and a shorter time commitment. One disadvantage is the increased cost in feed to meet the requirements needed for the pigs to grow properly. When pigs reach this stage it is important to raise them in the most economical way to derive the best profits. This will depend in large part on your goals and marketing strategy for your end products.

Another advantage to this phase is that you can enter the market at a time of your choice. This may allow you to take advantage of market forces, which might affect the other phases of pork production. By choosing your entrance time, you can also plan a time of exit that may fit your farm and family situation.

A wean-to-finish program allows the pork producer to concentrate on growing the young pigs as quickly to market weight as possible without maintaining a breeding sow herd.

This phase of raising pigs will also produce the largest volume of manure. The utilization and spreading of this manure will be a concern that you should consider prior to raising any pigs to market weight. Having an environmentally sound manure utilization program is as important as having a marketing strategy for your hogs.

PASTURE RAISING AND GRAZING

Unlike cattle, pigs cannot be raised solely on pastures or fed a grass-based diet. Pastures can provide a supplement to corn and soybeans and can make a significant contribution to and be an integral part of your swine operations.

Pastured pork is gaining in popularity, especially for those who want to farm on a small scale. This alternative approach requires less capital investment as the pasture is a major source of nutrients, particularly for gestating sows.

Pasture-raised pork programs are becoming more popular because some people believe the meat produced under these conditions has additional health benefits. Public interest in social and environmental issues is tied to how farm animals are being raised. Using pastures and grazing programs can reduce problems associated with animal-rights groups and other urbanites who view the treatment and care of farm animals as important issues. Pasture-raised pigs aren't confined in the way that pigs raised in large pork farms are. This can have a positive social impact on the community, especially with people who are troubled by modern pork-raising methods.

There are several advantages to raising pigs on pasture other than the social issues; these include lower initial investment costs, lower production costs, fewer facilities needed, and a sustainable method for producing pork. When pastured, pigs are less prone to cannibalism and fighting, perhaps because of the increased area they have in which to roam. There are fewer problems with waste management because the manure is quickly worked into the soil to inhibit runoff and lessen odors.

The therapeutic effect on pigs' immune systems is one positive aspect of raising them outdoors. Their immune systems develop due to their contact with soil, and being outdoors provides some isolation from other animals, which may aid in disease control. This has consistently resulted in healthier pigs with fewer respiratory diseases, rhinitis, and foot and leg problems. Healthier pigs need fewer therapeutic interventions, such as antibiotics, which also appeals to many consumers.

Not only is the open air better for the pigs; it's better for the people working with them. Raising pigs outdoors has a beneficial effect on the health of farm managers and workers because of decreases in dust, dirt, and odors.

There may be several disadvantages to pasture-raised pigs, including the possibility of problems with internal parasites because of the pigs' rooting behavior. There may be more labor involved with handling the pigs, providing access to water, or feeding them. If farrowed outdoors, more labor may be involved in providing good husbandry, and it will be more difficult to control their environment in extreme weather.

Feed costs are the largest production cost of any swine operation and will account for 55 to 85 percent of the total production costs. Providing good quality pasture for part of the year (or the entire year in areas that have mild climates) can significantly reduce feeding costs and increase profits. Quality pastures can provide needed nutrients for required growth and save on protein supplements and grain costs.

Pigs can eat vegetables, corn, or other crops after harvesting to supplement their diets. They can also serve as garbage disposals for food from the family table. This can have the double effect of providing the pigs with a source of nutrients, and increasing soil fertility due to their manure being deposited on the land.

Farrow-to-finish programs have the flexibility to ride out low market prices and capture high ones that may occur during the year. Although this option requires more management because of different sizes of animals involved, it has the potential to achieve a steady income.

Raising pigs to market weight requires more space per animal for movement and resting periods. Crowding pigs too closely is one cause of fighting, cannibalism, and tail biting.

The structures needed for raising pigs in open pastures are less expensive to construct than confinement buildings. Although good fencing is required, pigs can be pastured using a variety of methods, including electric fencing, which is easy to erect and can be easily moved from pasture to pasture.

Three important considerations for pigs on pasture are water, shade, and feed. While pigs may be able to grow on diets with a high percentage of pasture and grass and the conditions may be right for a little shade, they cannot survive long periods without water. Accommodations need to be made to supply water to pigs on pasture, whether it is by piping or hauling it to holding tanks in the field.

Permanent and Temporary Pastures

A permanent pasture is one that stays planted to grass or legumes or a mixture of each for several years in succession. Inter-seeding often occurs each year to replace plants that may have been rooted out or died out or to increase the plant population for the growing season to come.

Typically this seeding is done in the early spring so the plants can take root and make the best use of the growing season.

Temporary pastures are used for a short time and may remain as pasture for as long as there is feed available or it can be turned over for another crop. These are usually limited by the amount of forage they can provide, types of forage being grown, or proximity to buildings. Temporary pastures used for pigs may include sudan grass, sorghum-sudan crosses, and small, immature grains.

Legumes as a group have a higher protein, calcium, and carotene content than grasses and make excellent pastures. Legumes can also furnish an adequate supply of most vitamins, with the exception of D and B-12. Alfalfa, ladino clover, sweet clover, and red clover are legumes that may be used successfully for swine pastures. Ladino clover will not typically produce as much forage per acre as alfalfa, but it is higher in nutrition value. The use of ladino clover and alfalfa together can increase yields and improve the overall nutrient value of the pastures.

Perennial grasses, such as orchard grass, timothy, and brome grass, and certain fescues, such as endophyte-free tall fescue, can be used in mixtures with legumes to provide variety and an increase in sod formation. Grasses may help reduce legume loss during times of freezing, soil heaving, and stress.

An excellent ladino clover pasture will usually allow gestating sows to maintain their body weight even without additional feed. With average to excellent pastures, it is generally recommended that about two pounds of grain be fed daily for sows and three pounds for gilts on pastures. Sows and gilts should also have free-choice access to minerals and iodized salt during the first two-thirds of gestation. Lactating sows should receive two to three pounds of a 15 percent protein ration per one hundred pounds of body weight.

The location of permanent and temporary pastures may be important. You may designate different pastures for different management phases, such as keeping all dry sows in one area, farrowing sows in another, growing pigs in another, and so on. Planning ahead for these different phases may reduce the amount of work needed to set up your rotation.

Generally it is best if you can use pasture areas that are well drained and large enough to accommodate the size of your herd. The size of your herd is not as important as sound fences. Pastures should be well fenced either permanently or by electric wire to separate paddocks to prevent intermingling of different groups.

Some grazers use a wagon-wheel-type pasture arrangement, which may decrease labor requirements because the distance traveled when rotating pigs among paddocks is reduced. This design provides a central area for supplemental feeding and watering and reduces the need to place watering structures or feed troughs in each paddock. In these circumstances, sufficient width at the gates is needed to move equipment through for reseeding crops and moving pigs.

Pasture Health

Typically, well-grown pastures of moderate density or thickness will produce between 2,000 to 2,500 pounds of total dry matter per acre in the first eight inches of growth. This kind of growth under normal weather conditions will produce roughly five tons of forage per acre during the growing season.

Finishing pigs can be raised in several areas of the farm. They can grow well on large, open-air concrete platforms that are easy to clean, in pens that have a partial-solid floor area for resting and feeding and another area that is slatted where the manure can pass through to keep the pen clean, or on pastures with grain supplements.

Field peas can be grown in areas of your farm where pigs have access to pastures. Supplementing their diet with other nongrain sources of nutrients can help lower production costs while still maintaining good growth rates.

The pasture rejuvenation during the season will be determined when pigs are removed from a particular pasture. The longer they graze on the same pasture, the longer it will take to regrow the forage. Similarly, if they are not placed on a young pasture soon enough the grass and legumes may grow too quickly and become less palatable and lower in protein and nutrients. When this occurs it is best to mow the entire pasture and let it regrow on its own.

Sows have excellent grazing abilities and you can take advantage of this by grazing pastures. It will help lower your feed costs. However, not all pastures are suitable for sows. Pastures should be young, tender, high in protein, and low in fiber. Clovers and annual grasses, such as wheat, oats, rye, and ryegrass make excellent forages for sows during the cooler months of the year.

The balance between utilizing the young, tender grass and not overgrazing so the recovery time is extended is one thing that can be learned by experience. Having several pastures available will help minimize damage to one individual area that is continually used by your pigs. Too much time on one area will likely result in complete defoliation of the pasture and slow its recovery.

Pasture Stocking Rates

Stocking rates can be defined as the number of pigs that can be supported by the available acreage. These rates are easy to calculate and are important to prevent the overgrazing of pastures and still

maintain the viability and growth of the grasses and legumes. Pasture stocking rates will depend on several factors, including soil fertility, quality of pasture, size of pigs, and time of year.

Good quality pastures can contribute between 30 to 50 percent of the feed value of grain and supplement the needs of a gestating sow's diet. Because no single feed ingredient can provide all the nutrients pigs need daily, it is important to provide a nutritionally balanced diet for the least amount of money. The key to using different feeds is to make certain they are mixed correctly or used in appropriate amounts to provide a balanced diet based on the weight of the pigs and their stage of production.

Rotational grazing will help ensure that maximum use of your pastures is achieved by maintaining a young, tender stage of growth. Rotating your pastures will also help avoid excessive trampling and rooting of the plants. Recommended pasture stocking rates for sows with litters is between four to six sows per acre. You should be able to place between fifteen to thirty pigs, weighing up to one hundred pounds, per acre. From one hundred pounds to market weight you can put ten to twenty pigs per acre. For gestating or dry sows, you can place eight to twelve sows per acre.

Farmers who pasture their animals spend a lot of time studying their pastures and developing grass and legume stands to sufficiently feed their livestock. You should develop an understanding of pasture health, how to develop good pastures, and know

Wheat is a source of high protein that can be sown into pastures in late fall to grow in early spring as a supplement to the pigs' diet.

A pasture mix containing several different grasses is an excellent supplement for your breeding sows. Although pigs cannot be raised exclusively on forages, they can make up a significant part of the pigs' diet.

what kinds of yields to expect, because it will help you become better at raising pigs on pasture. You can develop a sustainable stocking rate by having a thorough understanding of the pasture's approximate forage yield and use this to decrease your feed costs.

Setting Up Paddocks

Pastures that are divided into separate feeding or grazing areas are called paddocks. They can be part of a larger field that has been separated by fencing or electric wire to control access. Paddocks may be located in several different places if your farm has areas that are not easily tilled for crops. You may be

able to develop several paddocks in one field to use in a rotational system.

The idea behind using paddocks in a rotational grazing program is to place the pigs in one pasture long enough for them to graze the grass down to between two to three inches before moving the animals to the next pasture. The time it takes for the pigs to graze the grass will depend on the time of year. Grasses typically grow quicker in spring and slower in summer and fall. The number and size of pigs on pasture will also be a factor, as are the weather conditions, because grass grows slower in dry versus wet conditions.

Unfortunately there is no set answer for your system. The best way to determine when to move your pigs from one area to another and still maintain good pasture growth is to walk through the pasture each day and determine how much they've grazed by the length of the grass. This is where experience will set you apart from other pig growers. You will learn to read the land and the grass in ways you may never have thought possible. Even experienced farmers continue to learn from their programs to get the optimum benefits from their pastures.

Pasture Housing

Depending on your climate, you may need to provide pasture shelters to allow the pigs to get out of the sun, rain, ice, snow, or other weather conditions. Some farmers use portable galvanized steel huts that can accommodate a number of sows or younger pigs at one time. Being able to move these huts from pasture to pasture or paddock to paddock will offer flexibility in their use and eliminate the need for permanent structures in each lot. Nonpermanent structures can also be used for farrowing areas, although individual draft-free shelters should be provided for each sow during farrowing. The amount of floor space available during farrowing becomes important because piglets can be crushed by the sow when there is not enough space.

Options for Small Producers

Pastures can provide excellent areas to raise pigs of almost any age and size and can help keep feed costs low. Public concerns about humane treatment of animals and food produced under natural conditions can be addressed with the use of pastures. Pastures allow pigs a chance to pursue their natural instincts, reduce their stress level with more space to freely move about, and increase their (and their handler's) level of comfort.

Adjoining pastures can be made into different areas to raise pigs outdoors. Having a rotation plan will help keep the different paddocks from being overgrazed or damaged by the rooting behavior of your mature pigs.

Portable huts are popular structures for pigs on pastures because of the shelter they provide for pigs to escape the elements, the space they offer for a large number of pigs, and the ease in which they can be filled with bedding or moved from one spot in the field to another.

A nonpermanent structure, such as this farrowing hut, can be used in many types of climates. The open rear end allows air movement in hot weather and can be closed during cold weather. Constructing them with sturdy attachments will allow easy movement to different areas of the pasture or to entirely different pastures.

CHAPTER 3
HOUSING AND FENCING

The type of swine operation you choose will largely determine the facilities you need. Conversely, the facilities at your disposal may have an important impact on how you structure your program. The building and facilities may already exist on your farm for the four major segments of a swine operation: farrowing and nursery for small pigs, growing, and finishing.

Indoor farrowing can be accomplished by constructing pens that provide enough space for the sow to lie down while giving birth, to move about, and where a barrier—such as a metal gate seen here—can be placed to give the piglets protection from her movements. Keeping the lights on for sows will allow them to see their piglets much easier and avoid stepping on them.

A structure for farrowing outdoors can be made of sturdy wood with an A-frame design. This allows the sow to claim the hut as her own at farrowing time.

Taking an inventory of your existing buildings will help you determine which building would be best suited for a particular phase of your swine operation. By incorporating ideas into buildings already on your farm, you can decrease your initial investment. Plans for building structures may be obtained from your county agricultural extension office.

Facilities do not need to be elaborate or expensive structures. If some buildings already exist on your farm, they may be converted into structures that will suit your needs. Important housing considerations are safety for the sow, her young, and their handler; adequate ventilation; convenience in cleaning out the bedding area; and sufficient warmth for the piglets during cool or cold weather. The area where your sows farrow may be located in an old barn, shed, or some other place where they will be separated from the rest of your pigs. Later the sow and her young can be moved together to another area of your farm.

The trend in pig production over the past several decades has been toward enclosed, environmentally controlled facilities, because they allow for the greatest production efficiency for large numbers of animals during all seasons of the year. As more families enter the swine production arena the production practices and goals vary. Trends include smaller herds and specialized goals, such as raising heritage breeds, developing niche markets for home-grown pork products, raising pigs in a more traditional way, and low-intensity pig housing. The structures you need may already exist on your farm, and with a little work you may be able to convert them into usable areas to raise pigs.

SHELTER DESIGN

Every shelter used for raising pigs should include three considerations: temperature, climate, and comfort. Fluctuations in any of these can reduce productivity and profitability as well as the general health of the animals.

Any designs for shelter or pens should take a pig's behavior into consideration. Pigs are smart and clean animals, contrary to most beliefs. They typically assign a certain function to different areas of their living space. If given a choice, they prefer a single sleeping area, which is generally the most comfortable area they can find, a separate area to deposit their manure, and another area for eating and drinking.

When constructing pens, consider these requirements in your designs. If there is room, most of their manure will be deposited in areas away from where they sleep. Place the water source and feeding area near the opposite end from their bedding. Water spilled on the floor or dirt will be a cold place to lie, and unless it is hot weather, they will stay away from the area. Locate the feeding trough or feeder a short distance away from the water supply and away from the sleeping area. If feed is located in the sleeping area, those lying down will restrict other pigs from getting access to the food. If feeders and water supplies are poorly situated within a pen, pigs will not establish separate sleeping or manuring areas and they will quickly become dirty and wet. Providing ample space is one important consideration when creating an area to raise pigs.

As the pigs grow, they will obviously need more room. Pigs weighing between 10 to 40 pounds typically require about four square feet per pig. They require up to seven square feet as they approach 150 pounds. Sows require about fifteen square feet on solid floors.

Pastures generally offer more room for pigs. If you pasture your sows there are guidelines that will help you assess your needs. For each sow or boar you pasture, provide a minimum of one-quarter acre. Each sow with a litter will need about two and a quarter to two and a half acres, while six to ten dry sows can be placed per acre.

Aggression between pigs is generally higher when many are confined to a small space. By providing adequate space for each pig, you will ensure the maximum amount of comfort and cooperation between them.

FARROWING

If you've decided on a farrow-to-finish operation, you should provide an adequate farrowing area. You will need an area where the sow can comfortably give birth to her young and where her piglets have adequate room to avoid her movements. The style of farrowing area may be determined by several factors, such as the amount of space available, your philosophy, the number of sows you will have farrowing at the same time, and the approximate farrowing times during the year.

It is important for all sows to have adequate space to farrow in order to lessen the loss of piglets.

You can have pens of various sizes to accommodate large or small sows to maximize the available space. Often a metal railing is attached to the lower sides of the pen to provide a space for piglets to escape their mother. At one end of the pen a barrier can be made that allows only the piglets to pass through so they can congregate under a heat lamp that is located away from their mother. Some producers use metal crates that are individual stalls where the sow is confined shortly before farrowing. She still can have her young and minimize the possibility of crushing or stepping on them.

The farrowing area should have electrical outlets nearby where heat lamps or heat pads can be used. This is important if the building is not a temperature-controlled environment. Pigs typically have a body temperature of 102 degrees Fahrenheit that needs to be maintained to prevent illness. Heat lamps can provide a supplemental heat source during cold weather. Lamps are placed over a spot where the piglets can congregate away from the mother and stay warm. Poorly installed or old electrical wires should be replaced to prevent fires or the overheating of wires.

There is a global trend today of government-issued guidelines that require pork producers to have pens for farrowing that provide sows with freedom of movement to turn around and allow them to express their natural behavior while still protecting their piglets. This pen has a nest area

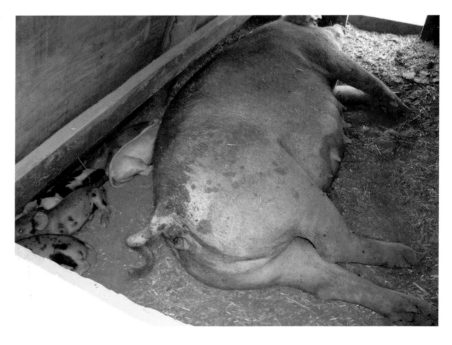

The interior of an outdoor farrowing hut can include a safety rail that keeps the sow away from the edges on the inside and gives the piglets room to lie down near her and avoid being crushed.

An open area in a barn can be divided to separate different sized pigs. Keeping groups of uniform sizes together minimizes stress and the dominance of larger pigs over smaller ones. Placing a straw bale in the pen gives the pigs an activity in pulling it apart while spreading the bedding around to lessen your work load.

covered with bedding but is separated by a barrier for the piglets. It has inward sloping sides that help the sow slowly lie down and provides a safe area for piglets. If this trend becomes part of the United States protocol for farrowing sows, you may want to consider building pens to meet those specifications. Designs for farrowing pens are generally available from your county agricultural extension office.

WEANING AND FEEDER PIGS

After they are born, young piglets quickly grow on milk from their mothers and the small amounts of feed that are slowly added to their diets. As they grow they can be moved to other areas. Young pigs are typically weaned from their mothers when they are three weeks old or weigh between 10 and 25 pounds. As they become less dependent upon their mother's milk, they start to eat feed supplements. Even when they are moved to another pen away from their mother, these young pigs are quite self-sufficient and simply need food and water in order to grow and thrive.

The facilities you have for newly weaned pigs can be very flexible. They may range from low-density housing units with group pens inside buildings to low-density housing with outside feeding areas or pastures and outside pig lots. The structures you use for this early growth period can also be used throughout their finishing stage if you do not have any other litters that will need the facilities.

Indoor farrowing units can use heating pads to replace heat lamps. They have the advantage of offering a consistently warm area without producing a draft. These pads can draw the piglets away from their mother when they are finished nursing and keep them warm.

Young pigs have a desire to stay warm even as they approach weaning. In the absence of heat lamps or pads, they may crowd one another. Because of their size and age, they are less likely to smother each other.

Finishing pigs for market requires more feed to reach target weights. While pasturing may supplement a diet, it cannot provide enough nutrition for fast weight gains. Large self-feeders, as shown here, hold a large volume of feed that is available at all times.

Some pork producers construct hoop structures to house growing pigs. Hoops or hoop structures are used to lower input costs. The hoop structure has one enclosed end and is open at the other, has wooden sidewalls with an arched tubular steel frame with a tarp stretched over the top as a cover, and is easy to move. Hoops typically have dirt floors with an outside area made of concrete or gravel for feeding and watering the animals. The dirt floor is usually covered with straw, corn stalks, or other suitable bedding material to make it easy to clean.

The relative low cost of a hoop structure makes them an attractive alternative to standard buildings. Depending on the size of your sow herd, one hoop structure may adequately house the number of pigs you plan to raise in a year.

GROWING AND FINISHING PIGS
During the growing and finishing phases, pigs eat as much as they can to reach a market weight between 230 to 280 pounds. Depending on your feeding program, this may take five to six months to accomplish.

As pigs approach market weight, they require more space. How much space you need depends upon the number of pigs grouped together and the size of the pen or pasture. Typically about eight square feet per animal is considered adequate for indoors, while a lot or pasture will generally provide adequate room.

Most problems occurring at this time tend to result from smaller-sized pigs being penned with larger ones. Keeping variations in size to a minimum will prevent many problems. Dominant pigs pick on smaller ones and can possibly cause injuries or poor growth rates.

Pigs at this stage grow best in temperatures between 60 to 70 degrees Fahrenheit. In winter, especially in colder climates, pigs need shelter from wind, rain, and snow. If they are housed indoors, they will need adequate ventilation to provide fresh air and prevent respiratory diseases. In summer, particularly during extreme hot periods, they will need to be kept cool either by using ventilation fans, shade trees, water sprinkling systems, or some other method to keep their body temperatures within an acceptable range. Because of the open air, odors will be less

A metal Quonset-style hut can provide a great amount of space for the young pigs and provide adequate shelter from all types of weather. When bedded with straw or other material, they make comfortable living quarters for pigs.

Larger hoop structures can be erected for use with a sow herd, weaning pigs, or finishing pigs. They can be used as temporary storage areas when not in use for livestock.

prevalent when hogs are raised outdoors. If they are raised indoors, with the use of adequate bedding the manure can be composted or hauled onto a field.

PASTURE STRUCTURES

If you plan to pasture your pigs there are structures that can be used to decrease the effects of temperature fluctuations and climate changes. These structures do not need to be expensive, but they must be large enough to provide adequate space for the number of pigs involved.

Outdoor housing requirements for bred sows or a boar are between 15 and 20 square feet. For a sow with a litter, the area needed is about 30 to 40 square feet. Shade in warmer months can be provided by trees, wooden structures, or a pole structure with a shaded roof made from various materials.

Some producers use an A-frame structure, which is easy to build and has the advantage of being easy to move around a pasture or other areas of the farm. These structures can be large enough to house a farrowing sow, shelter a sow with a litter, or serve as a shelter for a sow after the piglets are weaned. These simple structures can be used in many types of weather and climate conditions.

Water is essential for pigs. A water fountain located in a pasture area can provide an adequate supply that is easily accessible by all pigs. Placing it on a cement platform keeps the area from becoming muddy and dries quickly if water is spilled.

Positioning an outdoor water fountain away from other buildings encourages pigs to move about the pasture and still provides close proximity to a water supply. Placing it above the level of the surrounding area allows for adequate drainage.

A multi-use water tank can serve several needs. If you have cattle on your farm, pigs can drink from the side opening while cattle can drink from the top.

Abandoned buildings that sit on land for sale may find a new use after remodeling or renovation. Buildings that are still structurally sound can be converted in different ways to house pigs.

FENCING AND ENCLOSURES

Good fences are important when raising pigs, not only to mark property lines but also to humanely enclose the animals. A fence will keep your pigs in your pastures and your neighbors' animals out of your pastures.

If you own a farm, it is likely you already know the property lines or perimeter. If you are considering a farm purchase, there are several things you can do to ensure that you fully understand where the property lines are located. Knowing the exact boundaries of your farm is good business and will help avoid problems should questions arise if and when neighboring farms are sold.

BOUNDARY LINES

If you are working with a real estate agent make certain he or she knows exactly where the property lines are located. If the agent doesn't know or if the party you are purchasing the farm from can't tell you, it is worth the expense to have a surveyor establish all boundary lines around the farm before any purchase is made.

If the lines seem to be well established, such as the line fences being in place, you may feel reasonably satisfied that the land you are purchasing is the one described on the deed. In the case where a portion of the farm may have been sold in recent years, you may want assurance that the new lines are the boundaries for the property you are buying.

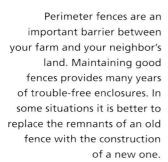

Perimeter fences are an important barrier between your farm and your neighbor's land. Maintaining good fences provides many years of trouble-free enclosures. In some situations it is better to replace the remnants of an old fence with the construction of a new one.

Some of the materials needed for constructing a good fence for pigs include T-posts, wire (woven and barbed), post hammer, shovel, a hand posthole digger, chain for pulling, claw hammer, and a manual wire-puller. Other items include pliers, wire cutter, tape measure, staples, smooth wire, wire clips, and a good pair of leather or heavy cloth gloves.

The construction of a new fence is best accomplished after the removal of the old one. If it is possible to remove the old fence, it will give you a clear and unencumbered area in which to work. Begin deconstruction by removing old clips attached to wire strands and any woven wire fencing. A pair of leather or heavy cloth gloves is essential to protect your hands from sharp barbs and wires.

In recent decades, many old fences have been removed to accommodate larger field equipment and to till the fence lines to keep weeds, small shrubs, and trees from growing. If fence lines are missing, it is important to know where they belong before constructing a fence of your own. Tearing down an improperly placed fence can be frustrating, time consuming, and costly. In this instance, it may be useful to hire a surveyor to identify the property line before you build a perimeter fence.

If you are buying a farm, ask to walk and view the fence lines with the owner or real estate agent to fully understand the boundaries. This walk can prove useful in a number of ways because it will allow you to see exactly what your prospective farm looks like. Walking the perimeter is vastly different from viewing it while riding in a vehicle or looking at a plat map, survey map, or photographs of the property. Walking the fence lines will also give you an opportunity to inspect their condition and to note problem areas that need attention.

Aerial maps of your farm can provide another perspective you can't get from simply walking your farm. Because they are taken from above your fields, aerial maps provide an overview of your farm not obtained from ground level. Aerial photography of your farm can be obtained from your county Farm Service Agency (FSA). Most, if not all, farming areas have aerial maps available and you can request a copy of your property.

With assistance from aerial maps, the relative boundaries of the fence lines can be seen. This map will show the farm's perimeter and give some indication of the lay of the land, which may be useful in handling any hills and valleys in the fence line.

Aerial photographs are also used to determine what areas of a farm are suited for different purposes. If you have multiple species of farm animals, the FSA office can assist you in identifying what areas are best suited for each species, such as for hay or permanent pastures, woodlands that may or may not be used for pasture, and areas suitable for cultivated crop production. These land assignments are made based on uses that will best protect the soil structure and return the greatest possible profit for each soil type.

FENCE PLANNING

If you plan on only raising pigs it is likely you will need two types of fences: permanent and temporary. Permanent boundary or division fences require different materials than fences for temporary lots. Whether you use only one type of fence or incorporate both will depend on if you graze your pigs in pastures in a rotation or if you plan to raise them in single lots. The rest of this chapter focuses on the construction of permanent and temporary fences that will allow you to provide safe and secure enclosures for your pigs. In either case, these physical barriers must provide sufficient strength to prevent or discourage animals from going over, under, or through the fence.

After the clips are removed, the wire should be pulled free from any posts or out of the fence line in preparation to rolling up the wire.

Another type of fence is a psychological barrier, which differs from a physical fence in that it discourages an animal from challenging a physical barrier of lesser strength because of the discomfort it will experience. This is a learned response, such as when electrified fences are used to contain animals.

Permanent fences are intended to last for many years with minimal repairs and should be constructed with sturdy materials to ensure long life. These are typically perimeter fences or fences along streams or waterways. Temporary fences are intended to last only a short time. They do not need to be stoutly constructed and are often made from less-expensive materials. Temporary fences are usually found within perimeter fences. If you plan to raise pigs on areas not suitable for other livestock or crops that are a distance from the farm buildings, you may want to construct one or more lanes, or pathways, for your pigs to gain access to those areas.

The key to a successfully built fence starts with a good plan on paper because alterations can be made quickly if you change parts of your plan. This allows you to calculate the length of the fence and the amount of fencing materials needed before you drive in a post. It will also help you estimate the cost of your fencing project.

In places where the wire is difficult to pull out by hand, mechanical assistance may be required. Pulling upward at the posts will extract both the post and any buried wire attached to it. This will allow the removal of any remaining wire clips.

Once the old wire has been completely removed from the fence line it can be rolled up and set off to the side. After it is hauled away it can be sold for scrap metal or recycled. Typically old wire is not used again for other fences. As you roll up the old wire, the steel T-posts can be set away from your work area and placed at the approximate distances that separated them before. This will allow those that are still good enough to use to be reset later.

Once your plan is developed on paper, you can walk out to your fields and see if your ideas will work or if you need to make changes before you start. It will probably be less expensive for you to construct your own fences rather than hiring someone to do the work. Although you can hire contractors who will quickly finish the project at a greater cost, with a little experience and guidance you can build a fence that will last a long time and withstand the pressures placed on them by your pigs.

The key rule to building a good fence is to build straight fence lines wherever possible. They are easier to construct, retain their tension for a longer period of time, and require fewer materials. You may need to make curved fences in certain areas, but try to avoid these whenever possible. They are difficult to construct and maintain.

Brush, tall grasses, or trees may pose a challenge and be more difficult to work with, but their removal will simplify the construction process and allow a clear path to place a new fence.

Permanent and Temporary Fences

One of the best ways to protect your livestock investment is to have a permanent fence that surrounds your farm. This applies whether you solely raise pigs or have other animals. Besides establishing a fixed property line, perimeter fences are the last line of defense if your pigs or other livestock escape from their designated pasture, feeding areas, or the small lots around your buildings. Good fencing protects and confines valuable animals by providing a barrier to restrict their movements.

Intact, well-constructed perimeter fences will prevent your animals from invading your neighbor's property, thereby relieving you of possible financial liabilities that may result from the destruction of property or crops. What animals your neighbors raise is beyond your control. Maintaining good fences for yourself will give you the same protection by keeping their animals out of your property.

With the wire, T-posts, brush, or other debris removed, you can scrape the fence line to provide a clean, clear area where you can lay down your new fence. If your fence work requires you to be on your neighbors' property, discuss the situation with them prior to doing any work. It is in the best interests of both parties to have good property line fences in place. Talking over your intentions may help avoid problems and misunderstandings.

Perimeter fences also protect your livestock by keeping them from wandering onto highways and roads, which could lead to collisions. If it is not possible to construct a permanent fence around the entire perimeter of your farm, consider building a fence in sections that will be most useful and provide the best barriers to enclose your pigs. It may be possible for you to complete the rest of the fence at a later date.

If there are areas you plan to use as pasture for several years in succession, you may want to consider constructing permanent fences around them. Farm ponds or other waterways should receive priority for permanent fences to control livestock access or only allow access for drinking. In areas where travel through waterways is necessary, access can be restricted with a permanent fence. You may also consider permanent fencing for fields where cultivated crops are grown and your pigs or other livestock are allowed access after harvest to root and graze the crop residue.

Temporary fences are used for a few days, weeks, or months and then can be removed. Movable fences are less expensive to build than permanent fences and are quicker to set up and take down. However, temporary fences are less durable and may not be as effective in controlling the movement of your pigs. Temporary fencing has the advantage of being quickly and easily moved from one area to another. It offers flexibility in providing pastures on different areas of your farm by allowing you to relocate the fence every year or several times within one year without constructing numerous permanent fences. Temporary fences allow you

Prior to digging any holes, determine the approximate distance between the two end posts by laying them out on the ground near the area to be set with the steel brace loosely placed into position. You will save yourself extra work by digging the holes as close to the desired spots as possible.

Manually digging holes to set end, corner, or line posts can be a challenging but rewarding job. One consequence of digging postholes by hand is to gain a better perspective of the soil structure of your farm. It may be possible to rent mechanical posthole drilling equipment and relieve much of the time and work required to achieve sufficient depths for your posts.

Having holes of sufficient size will allow you to move the posts in several directions as you line them up with each other and with the proposed fence line. A wider hole will also allow movement to attach the steel brace in the proper position.

to expand or contract the size of the pasture. This flexibility allows you to accommodate any increase or decrease in the number of pigs placed within a pasture, their relocation from pasture to pasture, and pasture rotations.

FENCING MATERIALS

Permanent fences can be constructed of several types of materials, such as wood, barbed wire, and woven wire. However, the most appropriate materials for swine fences are those of solid construction, such as woven wire, wire panel, and wood boards. In some cases, electric wire and barbed wire can be used in conjunction with those types of fences. High-tensile wire and cable wire fencing, unless electrified, is typically used for larger livestock, such as cattle.

Temporary fences are typically made from materials that can be easily and quickly moved, such as a single metal wire, poly wire, or wire tapes. These have the ability to be rolled up on a spool and moved from one location to another with minimal effort. Temporary fences are not considered for long-term use, although they can be used in that manner if needed. As a general rule, temporary fences should not be considered for or used in place of permanent perimeter fences, because they do not have the durability or strength.

The types of materials you use may have as much to do with your budget as any other factor. While there are differences in cost of materials available for fences it is usually wise to consider purchasing materials that will last a long time. wIf your plans and goals change during the course of your farming career, you can use your well-constructed fences for a variety of livestock without making major alterations or additions.

1. Prior to setting your posts, lay down and tightly stretch a single strand of wire in the fence line. This will help you position the posts before tamping them into the ground. Once the wire is stretched in the position of the old fence line, it will serve as a guide for setting the T-posts.

2. The stretched wire will help you accurately set the end post and the brace post. The bottom single strand of barbed wire serves as a deterrent to keep pigs from rooting under the woven-wire fence.

3. Use a retractable steel tape measure to determine the depth of the hole in relation to the length of the wood post. An eight-foot post should be set to a depth of three to three and a half feet to provide sufficient soil support and prevent heaving.

4. Straight posts make the strongest fence because they keep wires tight. Leaning posts lose wire tension. Use a level when setting corner or line posts to make sure they're straight.

While the cost of fencing materials is of great importance, you should consider balancing that cost with the performance and time element required. Although it may be a significant initial cost to construct a fence, this cost is spread out over the life of the fence. A well-constructed fence that requires minimal maintenance and repair over its lifetime is an excellent investment in the protection of your farm and livestock.

You can consider using a combination of fencing types if you are planning to construct fencing around your entire property. Pigs may require different fencing than other livestock species. Barbed wire may be used around areas that cattle use. Woven wire should be placed around areas used exclusively for pigs.

5. A long solid wood or metal handle can be used to tamp in the soil as you slowly refill the sides around the post. This compaction will help hold the post in place until the dirt settles over time.

Once the end or corner post is set and tamped in, attach the steel brace between it and the brace post, which has been partially set. Braces can be attached at an angle, such as this, where the load transfer is from the base of the brace post to the top of the end post.

The steel brace is held in place at each end by using large wood screws tightened with a box or ratchet wrench. It is preferable to use metal screws rather than nails at both ends because of their durability.

Woven-Wire Fences

Woven-wire fences make the best structures for swine lots, pastures, corrals, and areas where close confinement is needed for loading pigs onto trailers or trucks. Woven-wire fencing is strong and durable because it consists of a number of horizontal lines of smooth wire held apart by vertical wires called stays. The height needed to keep your pigs enclosed will likely be determined by the size of the pigs that will be in the pen. Because most pigs, especially very large ones, do not have the athletic ability to jump high, you may want to consider using short wire rolls with barbed-wire top lines to lower your costs rather than exclusively using tall rolls.

Woven wire is often more expensive than other types of fencing material because of the additional metal used in its production. Woven wire can be two to three times the cost of barbed wire for a similar distance covered and has approximately the same lifespan. Another factor to consider is that woven wire is sold in rolls, which are more difficult to handle than barbed-wire rolls because the woven roll is wider. More roll units are needed to complete a similar distance than with rolls of barbed wire.

Several types of woven wire are available with numerous combinations of wire sizes and spacing, number of line wires, and heights. The height of most woven-wire fencing materials ranges from 26 to 48 inches. The spacing between horizontal line wires may vary from 1½ inches at the bottom for small animals to 8 to 9 inches at the top for large

After the steel brace has been completely attached, finish tamping in the brace post. At this point, the post should be ready for you to attach the lower strand of barbed wire.

animals. Wire spacing usually increases with fence height. Similarly, the spacing between the stays can vary, and you should consider using woven wire with stay spacing of no more than 6 inches for piglets or other small animals. You can allow for 12 inches for large animals. Look for a manufacturer's label attached to the wire you are planning to buy. This label will tell you the type and style of the roll by using standard design numbers and will help ensure you are buying the correct fencing product for your needs.

Wood Fences

Wood boards are a strong and attractive building option for a fence. They are also safe for animals and those working with them. A significant disadvantage is that wood fences are expensive to build and maintain. It requires more labor to plan and build a wood fence than a wire one, although posts need to be set in either system.

Use several wood staples to hold the bottom wire tight. If a barb along the wire is available, pound one staple on each side of it. This will hold the wire tight and in place once you release the wire puller.

After releasing the tension from the wire puller, cut the strand but leave a sufficient length to wrap a portion around the post and tightly wind it around the stretched wire. This will aid in keeping the wire tight for a long period of time. Additional staples can be hammered around the wire on the back side of the post.

Board fences need to be attached to wood posts, but woven-wire fences can be attached to steel or wood posts depending on the purpose of the fence and preference of the farmer. The posts are typically set about 7½ feet apart to correspond with the length of the boards. The boards must overlap the wood post for the boards to be correctly nailed to the post. Spacing the edge of the wood posts exactly 8 feet apart will leave no length for the board to be nailed to the post.

When using boards for fencing, square wood posts work best because the board remains flat against the post and can be easily nailed to the post. If round wood posts are used, the surface area the flat board rests on is significantly decreased and will not provide as stable of an area to secure the board and nails.

Board fences usually are made from 1- to 2-inch-thick, 4- to 6-inch-wide, 8-foot-long boards. A wood board fence typically will be 4½ to 5 feet high. By spacing the boards 5 to 6 inches apart, you can calculate the amount of materials needed for your pig lot or pens.

If you are using wood boards for fencing it is best to consider placing the boards on the inside of the fence next to the pigs. For aesthetic reasons some prefer to place the boards on the outside of the posts, but this allows the pigs to push and rub against the boards, which will loosen the nails, and the boards may eventually fall off. While wood fences may be useful in small areas and around buildings, they also have a shorter life span because of weathering effects and often begin to splinter, break, or rot after a number of years. Other materials, such as hard plastic, that resemble wood fences, are available and may be useful for building fences.

Barbed-Wire Fences

Because it is less expensive then woven wire, barbed-wire fencing can be used for perimeters if your pigs do not have general access to those areas. Barbed-wire fencing can be used if you have large livestock, such as cattle, in your fields and makes it less expensive to fence the entire perimeter of your farm. A barbed-wire fence is a good barrier for pigs if it is only a temporary solution, such as allowing them to roam harvested fields in the fall or for preventing their access to highways, creeks, or woodlands where they would be difficult to retrieve.

In places where you feel the pigs may try to break through the barrier, you can install an electric wire. By running the wire at the pigs' eye level or roughly two-thirds of their height, it will discourage them from rooting around the lower wires of

a nonelectrified barbed-wire fence, which can possibly create loose wires or holes under the wire and provide an avenue of escape.

Be careful when working in close quarters near barbed-wire fences. The barbs can do the same damage to your skin as they can to pigs'. The nature of the fence is to provide discomfort from the barbs to keep the animals away from the fence. That very nature can also be the source of injuries to you or a family member if this type of fence is not treated with care.

Electric Fences

Electric fences are psychological barriers more than physical barriers because the only thing that would keep your pigs within a designated area by using a single wire stretched across a field is the shock the pig receives when it touches an electrified wire. The purpose of the shock is to scare or surprise the animal that touches it and alter the animal's behavior to avoid the wire. Although the initial shock may seem cruel, it is an effective and humane management tool for use with animals.

Electric fences offer flexibility and can be used for temporary fencing, although they can be useful in conjunction with permanent fences. Used as a movable or temporary fence, an electric fence can be made with one or two strands of smooth wire or a poly tape that has small metal wires woven into it. The poly tape is more flexible and easier to handle and move from one location to another. In either

Use smooth wire and attach it to opposite ends of the end and brace posts for additional support and to keep the posts from shifting away from each other over time. The wire can be twisted tight by using metal pipes or bars.

The stretched wire along your property line will help you place any line posts correctly as support for your fence. Lifting the wire with your fencing pliers will help you safely position the wire. Laying the wire first will help you find the correct line position to dig holes for your support posts.

case, the wire is energized by an electric controller that receives its source from a standard farm electrical outlet or a solar-powered pack.

The electricity used in the wire is typically harmless to an animal or human, although it may cause momentary discomfort. A controller or charger regulates the flow of energy by supplying short pulses of electricity that travel along the wire. When your pig comes in contact with the wire, it will complete the circuit from the wire through its body and through the ground. This discomfort will discourage further contact with the fence. Typically an electric fence is not continuously electrified but remains effective because of short electrical bursts. In many states it is unlawful to use any electric fence unless a controlling device regulates the charge on the fence wire.

Solar power packs rely on the same principle as electric control devices but have the advantage of not using an electric source for which you

pay. Solar packs derive their energy from the sun and deliver the current through the wire. Once fully charged, some packs have the capability of providing a low impedance current without the sun for up to two weeks. These may be useful in areas that are a distance from your buildings. Battery-powered charges are also an option, but they tend to have a short lifespan before they have to be recharged or replaced.

Electric fences have several advantages, including easy movement from site to site and less bracing for corner or end posts. The effectiveness of an electric fence diminishes when the electric current no longer runs the length of the wire or when vegetation grounds the wire. In order to keep your electric fence in top condition, periodically check along the length of the fence to make sure the current is traveling through the entire fence. Also keep the grass underneath the fence trimmed so that no vegetation touches the fence to ground the current.

Like other animals, pigs require training when they first encounter electric fences and often will not be aware of the fence until they come in contact with it and receive a shock from the wire. It is important for the fence charger or solar pack to be properly maintained for the temporary electric wire to be effective in keeping pigs inside the fenced area.

It is highly recommended that you do not use homemade or inexpensive, high impedance controllers. These devices may cause serious injury or death to animals or humans, and the use of poorly designed controllers may result in grass fires along the fences. Under no circumstances should your electric fence be directly connected to a live electrical outlet or high-voltage electric source.

High-Tensile and Cable-Wire Fences

High-tensile-wire and cable-wire fences work well for cattle confinement areas, such as corrals, around barnyards, and in loading areas, but they do not work as well for pigs. Both are smooth materials anchored to posts and have the ability to absorb shock from animal movement without causing injury. Because high-tensile wire can be electrified, it may have some use in your facilities, but since these types of materials are typically used in confined areas, it may not suit your purposes, especially if humans are going to be in close contact with the fence. In that instance, the electricity must be switched off and the effectiveness of the wire is lost. The same basic principle holds for cable wire, although it is not typically hooked to an electric source.

Wire Panels

Alternatives to wood fences are wire panels. Although similar in construction to woven wire, these panels are made from heavier metal material and are welded at the joints to provide a sturdy, long-lasting fencing option. Wire panels are similar to woven wire because they can be used in close confinement areas, such as lots, small pastures, loading areas, and around buildings.

Wire panels can be constructed as sturdy alleyways between buildings and are easily erected and quickly taken down. They are useful when moving pigs from one area to another if a trailer or other moving vehicle is not available. There are many and varied uses of wire panels that provide the flexibility of a fence combined with portability.

FENCING TOOLS AND POSTS

Heavy cloth or leather gloves are the most important items you will need when constructing fences. They are absolutely essential when working with wire as they reduce the chance for injury and severe cuts to your hands and arms, especially if you are constructing a barbed-wire fence.

Some of the tools you will need to construct a fence include fencing pliers, posthole digger, protective eyewear, tape measure or reel, and wire puller. When you have determined the type of wire that will work best for your situation and budget, you will also need staples, wood posts or steel T-posts, and clips.

You can purchase many of these items from farm supply stores, lumberyards, farm catalogs, hardware stores, or Internet companies. Typically when purchasing a large quantity of wire fencing, boards, posts, braces, and other heavy items, it is less expensive to find a local source. Shipping costs can quickly mount from other sources unless they are included in the total price. Be certain to fully understand any shipping costs involved whether or not you purchase from a local business. Sometimes you can receive a discount on a large volume purchase and for the hauling of these materials to your farm if you can't do it yourself.

If you are using woven- or barbed-wire materials, you should understand that different classes of zinc coating have been established. Generally speaking, the higher the class number, the greater the thickness of the zinc coating, which leads to a longer life of the material. Dealers in wire fencing can offer advice on the type and gauge of wire that will best suit your needs.

Using a steel post hammer is the simplest way of pounding the T-posts into the new fence line. By using the wire as a guide, you can keep your T-posts in a straight line.

If a part of the original fence does not need to be replaced, the old portion can be attached to your end post before you attach the new wire.

Positioning all T-posts an equal distance apart between the wood line posts will help solidify the strength of your fence. With the end and brace posts set and the wood and T-posts in place, you are now ready to roll out your woven wire and attach it to the T-posts with wire clips. Your line fence is now properly built and will provide you with years of service.

Fence posts are used to keep wires apart and in place and to provide stability and strength for the fence line. Posts are commonly made from wood or steel, and each type has its advantages and disadvantages. Wood posts have an advantage over steel posts in strength and resistance to bending. Permanent fences often require decay-resistant fence posts. Pressure-treated wood posts can last as long as steel posts. Wood posts come in different sizes so it is important that you use larger posts for the corners and braces, while small ones can be used as line posts.

Wood posts need to be long enough to support the fence height and the depth they are placed into the ground. A satisfactory post length is 6 inches more than the distance from the height of the top wire above the ground to the depth of the post in the ground. For example, say the top wire is to be 48 inches above ground level and you set the post 3½ feet deep in the ground. This equals 4 feet plus 3½ feet plus 6 inches, or 8 feet for the length of the post.

The advantages of steel posts are they cost less than a similar wood post, can be driven into the ground more easily without the need for digging holes, and they weigh less for easier handling. Steel T-posts generally are 5 to 6 feet in length.

You can determine the approximate amount of fencing materials you need by calculating the distances involved. If you are constructing line fences, use a reel tape to measure the distance as you roll it over the area you walk. Short distances can be measured by a standard roll-out tape.

A woven-wire fence will provide a long-term barrier for your pigs. It will keep them safe and within the boundaries of your farm.

Wire panels are an effective material for use in small lots. The smooth wires are sturdy but safe because there are no sharp points to cause injury to you or your pigs.

Electric fences can be set at various distances above the ground to deter small or large pigs. Pigs that have been trained to avoid these wires are easier to keep within their designated areas.

A lane that serves several pastures can be constructed using wire panels and lined with crushed rock to prevent ruts or gullies. A well-constructed lane will eliminate escapes, provide easy access in wet weather, and retain the integrity of the soil structure.

Solid end or corner posts, tight wires, and using the right materials are the main ingredients to building a permanent fence with a long lifespan. Every fencing job has different requirements and each fence presents a slightly different approach than another. Like other construction and maintenance jobs around the farm, building a fence requires proper techniques and a common-sense approach.

FENCE CONSTRUCTION

Building a new fence is easier if you first clear away any old fence materials or vegetation that may have grown in the fence line. This will leave you a clear path in which to work. If this is not possible, you should at least build the fence line away from obstacles.

By locating the corner or end posts as your first task, you will be able to easily plan your fence layout, because you will be aiming for those corners as you unroll wire. At the same time you identify these end points, you will need to determine where the gate openings or permanent lanes, if needed, should be located.

You can build a good fence by understanding the importance of solid corner posts and the tightness of wire. To begin, lay out the fence line by first locating the corners and gate openings.

Set the corner and brace posts first, as this will give you a solid base to attach the wire puller to stretch the wire you have rolled out along the fence line. The posts and braces are similar in importance to the foundation stone of a building. Having solid corner and brace posts will make the initial tightening of the wires easier and will help hold the wire tension for many years. The corner posts can be set by driving them into the ground with a mechanical post driver, digging a hole and tamping around the post with dirt, or cementing them into the ground. A depth of 3½ to 4 feet is recommended for corner or brace posts because of the tension exerted by the stretched wire. The shallower the depth, the more likely the corner post will eventually work its way out of the ground. A post moving vertically, even 2 inches, will provide enough decrease in wire tension that the fence will become slack, eventually provide gaps in the wire, and possibly allow your pigs an opportunity to push through it.

With sufficient depth, the corner and brace posts will not be affected by frost upheaval in colder climates. Depending on the soil type, the depth of the post may need to be increased and anchors put into the bottom to keep it from slowly working its way to the surface. Rising posts usually occur in mucky or marshy soil types where the ground surface may not be as solid or stable as in other areas of the farm. While it may take more time to anchor these kinds of posts, the result will be worth the extra effort.

After setting the corner post, the brace posts can be placed. The brace post provides support to the corner post because it transfers the pulling force of the wire from the corner to the brace. This is important, because when a wire is first stretched, the pulling force on the corner or end post may reach 3,000 pounds. Cold winter temperatures can cause the wire to contract, which can increase the pull to 4,500 pounds. The corner and end post assemblies must be strong enough to withstand these forces or the posts will slowly pull out of the ground and the wire will lose tension.

You can set the brace posts by using the guide wire previously used to line up the fence, or it may be just as easy to visually sight them. The distance between the brace and the corner post should be two and a half times the height of the fence for maximum support.

There are several ways of setting the line posts in the fence line construction. Depending on the length of your fence, the spacing between wood posts is normally 20 to 25 feet. With steel posts the spacing is normally 10 to 15 feet.

If the total length of the fence is greater than 650 feet between corner posts, it is advisable to insert a braced line post every 650 feet. This will help strengthen the fence and maintain its tension over long distances. Braced line posts are also useful when the fence is built over rolling land or hills.

Once the layout for the perimeter fence is finished, you can plan for any interior fences, corrals, waterway barriers, or other enclosures.

Gates and passageways for your pigs should be located in the corners of the fields nearest to the farm buildings. Having the entrances located in corners helps with the movement of pigs from one field to another and allows you to have your corner posts and gates in areas that do not break up the fence line.

A lane should be located in the driest area possible, such as along a natural ridge or some other higher land feature. Constant movement of animals within lanes can eventually develop gullies. If well-drained areas are not available for your lanes, consider using a lane fence that can be moved after several years, or else fill the lane with gravel for better footing for you and your pigs. If possible avoid placing lanes through wet or low-lying areas, because the animal traffic will soon turn that area to deep mud as the pigs will find it a comfortable and convenient place to lie down, particularly in hot weather.

HELP AND ADVICE

Fence construction booklets are available from county agricultural extension offices or fence manufacturing companies. Although there are initial costs involved in constructing a fence, there may be programs available through land conservation agencies that can help offset some of these costs when applied to certain management practices. Check with your county agriculture extension office to learn more about these programs.

Pastures can be divided off a lane to allow for easy movement from one area to another. Developing a fencing plan on paper will make the construction of your paddocks easier.

CHAPTER 4
FEED AND MANURE

FEEDING FOR PROFIT AND PRODUCTION

Pigs have a monogastric digestive system, which is a single-chambered stomach that breaks down and processes food into usable nutrients. The monogastric digestive system is simple compared with the complex four-chambered stomach of a ruminant, such as a cow. The monogastric system uses a simple process where the food enters the mouth and is first broken down by mastication, which moistens and lubricates the food before it is swallowed and passed down the esophagus to enter the stomach.

Enzymes are secreted inside the stomach and mixed with the food to further break down the particles into sugars, carbohydrates, proteins, and minerals. As these processed substances are passed to the large intestine, nutrients are absorbed and duodenal cells secrete hormones that stimulate release of bile and pancreatic enzymes. This mixture is passed to the small intestine where water is absorbed, and the remaining unused material is passed out as feces.

Most of the feed eaten by young pigs is used for growth and maintenance, while mature pigs primarily use feed for maintenance and reproduction. Digested feed that is not needed for growth, maintenance, or reproduction is stored as body fat or passed out of the body as waste products.

NUTRIENT VALUES

Nutrients used to feed your pigs can come from a variety of plant sources and be processed in several forms. Good sources of nutrients in sufficient quantities that are consumed daily are essential for all ages of pigs, from newborn to market weight and the breeding herd. These nutrients will support and enhance several different body functions: growth, reproduction, maintenance, milk production, and fattening. Proteins, carbohydrates, fats, vitamins, and minerals will promote and supply the building blocks for growth and increase the size, length, and mass of the bones, muscles, and skin cells.

Reproduction is enhanced and supported with a well-balanced diet from pre-breeding to the development and growth of the unborn pigs from mating to gestation. Nutrients supply the energy for maintenance, normal daily activities, and to repair and replace worn body tissues. For pregnant and nursing gilts and sows, nutrients help produce milk for the young. Finally, nutrients are essential to the formation of fat tissue under the skin, around the muscles, and in the body cavity to enhance the flavor of the meat.

NUTRIENT SOURCES

There are a wide variety of plants that can provide the essential nutrients needed for profitable growth and maintenance. Rations—the formulated daily diet of pigs—are closely tailored to provide optimum health and growth at each stage of a pig's life. Using available feedstuffs that can promote excellent growth while remaining low in cost is one key to profitable pork production. This should become part of your plans and goals. You can use a variety of feed sources to raise your pigs that fit your farm situation and cropping plan.

Corn and soybeans are the most common sources of carbohydrates and proteins supplied to the diets of pigs and are generally available in sufficient quantities in all areas of the country. Whether you raise corn and soybeans on your farm or purchase them from an outside source may depend on your goals, number of potential acres available for crops, equipment on hand, and perhaps your expertise and desire to raise crops. If

Corn is a member of the grass family, but it is an unusual one. It belongs to the group of six true grains, or cereals, which also includes wheat, barley, oats, rice, and rye. Corn typically serves as the basis for most pig rations.

TYPICAL ENERGY SOURCES FOR PIGS

Grain	Feed Value	Maximum Ration Content for Sow and Pigs
Corn	100%	100%
Wheat	95%	50-75%
Milo	95%	50-75%
Barley	90%	25-50%
Oats	80%	25-50%

This chart shows the typical energy sources for feeding pigs. There are others, but these five main grains can be easily grown in most parts of the United States.

raising corn and soybeans is part of your farming plan, you may have an adequate amount available to supply the yearly requirements of your swine operation. If you do not raise crops, you will have to find a source for protein and carbohydrates, such as a local mill, feed supply companies, or neighboring farms. Other potential energy sources for your swine rations are cereal grains, such as wheat, oats, barley, and milo, or grain sorghum.

These crops may either be raised or purchased, depending on your situation.

The amount of nutrients needed in a ration is determined by several factors, including age and weight of the pigs. Rations are often adjusted to account for these changes. The ration is normally changed to provide more energy and less protein as the pig grows. The goal is to maximize growth at different stages of life with the available feedstuffs.

Soybeans are a source of high protein for feeding your pigs. Whether you raise soybeans yourself or purchase them may be influenced by factors such as the amount of acreage you have available for crops.

The protein level requirements differ for various sizes of pigs. This chart gives examples of the levels needed by different classes of pigs.

PROTEIN LEVELS REQUIRED BY SWINE

RATION	PROTEIN %
Creep Feeding	18-20%
Growing (50-125 lbs.)	15-16%
Finishing (125-240 lbs.)	13-14%
Young gilts & boars	15-16%
Older sows & boars	13-14%

Traditional sources of nutrients, such as grasses and legumes, have seen a resurgence in popularity, particularly among beginning pig farmers and those desiring a different production protocol. In past decades, pastures and grass-based pork-production programs diminished in use as confinement systems increased. The use of grasses and legumes in raising pigs is a viable alternative to a diet consisting completely of corn and soybean products.

PROTEIN AND ENERGY REQUIREMENTS

The amount of protein needed in a ration and the quality (referring to a balance of amino acids) of the proteins are extremely important. Protein is required for the growth of muscles and organs and the production of enzymes, hormones, and milk. Cereal grains, oilseed meals, and animal byproducts can meet your pig's protein requirements.

Cereal grains, such as wheat, barley, milo, and oats, are often too low in protein to be used alone. They may be mixed with oilseed meals, such as soybean meal or canola meal, as supplements to other protein sources. Animal byproducts, such as fish meal, bone meal, dried skim milk, and whey products, may also be used as high-quality protein sources, but they are generally very expensive.

Wheat is a cereal grain that is too low in protein to be used alone, but when mixed with oil seed meals, including soybean meal or canola meal, it can become a significant part of a pig ration.

All grains vary in protein content depending on the conditions under which they were grown and occasionally the specific variety planted. To best utilize the feeds you have available, or if feeds are purchased, it is critical to have them analyzed for nutrient content. This will allow you to formulate rations that do not waste nutrients exceeding the requirements of your pigs or to underfeed needed nutrients. When purchasing feed ingredients, compare prices of the various grains and sources of protein and use the lowest appropriate cost combination in the mix.

Energy can be supplied as carbohydrates, fats, or proteins. Fats contain about 2.25 times as much energy per unit of weight as carbohydrates and proteins. Typically protein supplements are not used as energy sources because of their increased cost. In some instances, you can substitute fats for carbohydrates to increase the energy density of a ration. This may be beneficial during hot weather, when pigs naturally eat less feed. Higher energy rations will help keep the energy intake the same even when less feed is consumed.

Energy is available to the animal when it breaks down the grain kernels. Soft-coated grains, such as wheat, barley, and oats, are easily digested by pigs.

Corn, however, has a hard outer coating containing cellulose that is difficult to break down. Corn kernels need to be cracked or ground to break the cellulose coating and allow the pig's digestive system access to the germ and endosperm to break down the sugar, starch, vitamins, and minerals.

Carbohydrates supply sugars that are metabolized and used as energy for the formation of new tissues, including muscle and fat, and for milk production during lactation. Surplus energy fed to the pig will be stored as fat, which can be metabolized at a later time for additional energy.

Typical Protein Sources

Pigs do not specifically need protein, but they require amino acids for the formation of muscle and other body proteins. Twenty amino acids are needed for proper pig growth, with ten required as dietary essentials: arginine, histidine, isoleucine, leucine, lysine, methionine, phenylalanine, threonine, tryptophan, and valine.

There are many grain and plant byproducts that can be used in pig diets to provide protein and amino acids. Commonly used plant byproducts include corn gluten meal, hominy feed, brewer's products, distiller's grains, wheat, bran, and alfalfa meal.

A feed bin can be constructed near the pens where the weaned or young finishing pigs are housed but won't have access to it. Ground feed can easily be carried to the pigs.

MINERALS AND VITAMINS

Many metabolic processes in pigs depend on small quantities of minerals and vitamins. These nutrients are important because they aid in the development of strong bones and teeth and are required for muscle contraction, hormone function, and blood clotting. Failure to provide sufficient amounts of minerals and vitamins, either through feedstuffs or commercial supplements, can severely affect the growth and health of your pigs.

Mineral and vitamin requirements can be supplied by homegrown feeds. If you rely solely upon these feeds, it is wise to have them tested for their nutrient content to be sure they are supplying the required amounts of minerals and vitamins.

More than fifteen minerals have been identified as essential for pig growth. Three of them—calcium, phosphorus, and iodized salt—combine for the largest requirements. Other important minerals, such as iron, copper, zinc, and manganese, can be supplied in a trace-mineralized salt combination.

Mineral and vitamin premixes can be purchased to supplement homegrown feeds, but they can be very expensive. The amounts of minerals and vitamins required will change as the pig grows and matures. While minimum mineral amounts are essential for proper growth, excessive amounts of certain minerals, such as selenium and fluorine, can be detrimental or even toxic to pigs. Understanding the requirements for minerals and not exceeding them will prevent problems.

Vitamins regulate many body functions and are essential for proper growth. Vitamin A affects the eyes and vision. Vitamin D affects calcium and phosphorus metabolism, and B vitamins support the nervous system. Some vitamins can come directly

APPROXIMATE WATER AND FEED CONSUMPTION			
Pig Weight (lbs.)	# Dry Feed/day	# Water/day	Gallons/day/approx.
50	4	8-12	1-1.5
100	6	12-18	1.5-2.2
150	8	16-24	2.0-3.0
200	10	20-30	2.5-3.5
250	12	24-36	3.0-3.5

This chart shows the approximate water needs for feed consumption by pigs. These amounts may vary depending on time of year, temperatures, and amount of feed consumed.

If you plan to purchase most of the grain for feeding, a large upright metal holding tank may be useful. These hold several tons of feed at one time, and buying in bulk may be a cost-saving consideration in your area or farm situation.

The simple design of a trough feeder keeps feed waste to a minimum and allows smaller pigs to eat outdoors. Typically these types of feeders are replenished daily with either solid feed or water.

An upright metal feeder can be used in areas where finishing pigs are housed. They are durable structures that hold a large volume of feed. If outdoor feeders are used, they should be covered to keep out rain or snow and prevent feed spoilage.

Water is essential to a pig's diet. Many types of water fountains are available, including those that have heating capabilities. A heated fountain allows pigs access to water in cold weather. When placed on a concrete platform, a water fountain can provide long service with minimal maintenance.

from the sun, alfalfa meal, green pasture, or synthetic premixes. Animals confined in buildings must be fed vitamins because of their lack of access to sunlight.

PROVIDE WATER

Water is the most important nutrient of your swine operation. It is needed to maintain body temperature, to carry digested food through the body, and to remove body wastes.

Water consumption varies with changes in temperature and feed consumption. A pig will typically consume two to three times as much water as its daily feed consumption, regardless of its size. This consumption can translate to between 1½ to 2 gallons of water a day over a six-month period. Having adequate supplies of clean, fresh water is essential.

FEED ADDITIVES

Feed additives have been used in swine rations to improve daily weight gain and feed efficiency. Depending on your goals and the type of pig-raising system you pursue, these additives may not be attractive ingredients for your program. Feed additives have generally included antibiotics, and

arsenicals and sulfa compounds have been used in combination with antibiotics to promote growth. Sometimes additives have been used in preventative programs, but the routine usage of antibiotics has been open for debate in recent years.

FORAGES FOR PIGS

Traditionally pastures formed an essential part of a successful swine operation. They provide minerals, vitamins, proteins, and other essential nutrients that lower feeding costs but still allow acceptable growth rates. Pastures also provide a means for exercise and generally require less labor for the farmer in handling manure. Pastures often have been replaced in recent years with confinement systems that improve rates of gain and provide a simple means of disease and parasite control.

With developing niche markets for pork products raised in a more natural environment, the use of pastures and other forages by pork producers has increased. Although a pig's monogastric digestive system does not lend itself to using great quantities of pasture or roughage, pasturing can provide quality protein and certain vitamins and reduce the total feed requirements from supplements and grains.

Oats are another cereal grain that can be grown on most farms and used as a supplement to pig rations. Oats are lower in feed value as an energy source than other grains and should only make up one-quarter to one-half of the ration content.

If pastures have mature growth, it is advisable to cut the old growth and harvest it to be used as supplemental feed or bedding. Cutting the old pasture growth allows the plants to regenerate for new growth.

Generally these savings can amount to 3 to 10 percent of the grain and as much as half of the protein needed for growing and finishing pigs, depending on the type of pasture, age of the pigs, and your management system. Good pastures containing alfalfa, ladino clover, and grasses can help lower feed costs, help maintain high-level reproductive capacity of boars, and increase litter size when compared to confined animals. For these reasons you may wish to consider pasturing pigs on your farm. Legume and grass pastures can provide supplemental protein, requiring less purchased feeds and about half as much grain as pigs raised in dry lots.

After wheat or oats ripen, they are harvested and the grain is used for feed or can be sold. The residue left behind is referred to as straw and can be baled in either large or small square bales or round bales. Wheat or oat straw makes excellent bedding material for all sizes of pigs whether it is left as long stem or is chopped into smaller pieces.

These are four sample rations that can be used for different feeding groups of pigs. They are based on a one-ton mix, which means the total amount of the different feedstuffs equals 2,000 pounds. Rations can be altered to accommodate changes in feedstuff availability, costs, and other factors.

A TYPICAL MIXED RATION—BASED ON A 1-TON MIX—FOR DIFFERENT GROWING GROUPS

Baby Pig-20% Protein
1000# Yellow Corn
200# Rolled Oats
250# Dried Whey
600# Soybean Meal
50-100# Vitamin-Minerals

Young Pig-18% (10-50 lbs.)
900# Yellow Corn
200 # Rolled Oats
250 # Dried Whey
500# Soybean Meal
50-100# Vitamin-Minerals

Growing Pig-16% (50-125 lbs.)
1500# Yellow Corn
400# Soybean Meal
50# Vitamin-Minerals

Finishing Pig-14% (125-250 lbs.)
1600# Yellow Corn
350# Soybean Meal
50# Vitamin-Minerals

Pastures are highly recommended for the breeding herd because they provide exercise and many nutrients needed by sows. Pastures can be seeded in areas not easily adapted to cropping or where mechanical harvesting is difficult or impossible. Manure handling and management is easier because pigs spread their waste as they move about the pasture. Some pig growers assert that pigs raised on pasture are healthier, because their immune systems are better developed and therefore stronger, and more resistant to diseases common to pigs raised in confinement.

The pasture system may have some disadvantages, including more labor required for handling, feeding, and watering the pigs. Raising hogs on pasture may decrease the amount of cropland available, with less control over extreme temperatures or internal parasites. These disadvantages may be mitigated with your management practices, thereby making pasture raising a viable option for your farm.

FORMULATING RATIONS

Ration formulation is possible with many computer programs available today. These programs can determine the best combination of ingredients to provide needed protein, energy, minerals, and vitamins at the most economical price based on the nutrient levels in available feedstuffs. For proper calculations with homegrown feeds, testing the feedstuffs available on your farm will help develop rations that meet your pigs' requirements and minimize outside purchases.

If you purchase all of your feed, these programs can help formulate rations based on products readily available in your region and calculate the least-cost option for your program. Most feed companies can help develop a ration for your pigs. These rations usually come with their products being recommended, which may or may not be to your advantage. County agricultural extension agents have access to ration formulation programs that may be helpful to you, and they can offer advice on where to seek additional help.

Pigs seek cool areas in hot weather, and wallows will develop over time as they lie in them to find relief from the heat. While these can't be entirely eliminated in pasture situations, they can be kept to a minimum by rotating pastures, keeping water sources on concrete areas, or providing shelters in the field with enough space to house the number of pigs you have.

MANAGING MANURE

Pork producers have an obligation to operate their farms, regardless of the size, in an environmentally responsible manner. Knowing and understanding the options available for manure usage will allow you to handle it in a way that protects the environment and enhances the productivity of your farm. Manure storage, handling, and usage are not complicated procedures that require expensive facilities. Depending upon the number of animals you decide to raise, you can develop a nutrient management plan before any animals arrive on your farm. Planning ahead will make the utilization of manures more effective in plant nutrition and in lowering your crop fertilizer costs.

Nutrient management is the handling and effective use of the manure produced by the animals on your farm. Manure is an unavoidable byproduct of pigs. How it is used will largely determine if it is a valuable asset to your farm or a costly liability. The purpose of a nutrient management program is to use the manure components in a way that provides nutrients for your soil through balanced applications across your fields (or those of your neighbors), while minimizing any detrimental effects on the surrounding environment, such as waterways, streams, or other water sources, including wells.

As farms have become larger, nutrient management has become a priority issue with farmers, environmentalists, government agencies assigned the task of maintaining clean water resources, and the public at large. Pork production can impact water quality by contaminating surface waters or groundwater. The contamination is the result of nutrient runoff when the manure is spread on the top of fields. Runoff most often occurs on frozen ground where the manure has little or no chance of being absorbed into the ground prior to rain or snow melt. Farm waste runoff into streams, creeks, lakes, and other water bodies has become one of the most contentious issues between rural landowners and urban populations that view these waterways as recreational areas. Well-planned use of manure can avert problems before they occur and can be a valuable asset instead of a liability to your farm.

Natural habitats are enhanced by the proper use of nutrients from farms. Everyone has the responsibility to safeguard our natural resources for future generations.

An even application of solid manure ensures that a wide area of your field is covered. A tractor-drawn manure spreader such as this one can spread solid manure in even patterns across a field. Proper use of manures can reduce outside input costs. Solid manure is an excellent source of fiber and organic matter, helps replenish lost nutrients, and loosens soil particles for better water absorption.

MANURE USES

Normal digestive processes of a pig's gastrointestinal system produce fecal material and urine that is expelled as manure. Manure quickly decomposes under warm, moist soil conditions and releases nitrogen, phosphorus, potassium, and other nutrients into the soil.

Early civilizations recognized the benefits of manure on their crops and plants and used it for fuel, shelter construction, and sport. When commercial fertilizers became available, manures became a disposal problem rather than an asset. Today, farmers try to keep manure runoff from reaching water sources and use the nutrients contained in manure to improve soil fertility, structure, and composition. Field plants absorb nutrients from the soil in order to grow. Replenishing these nutrients with manure can lower commercial fertilizer purchases.

When mixed with bedding, such as straw or hay, solid manure can add fiber and organic matter back into the soil. This combination, when plowed or stirred back into the field, can loosen soil particles and make the soil more porous, which allows it to both absorb and retain more moisture. Loosening the soil particles relieves compaction and will provide the plants with a soil structure that allows a better root system to develop for better growth.

MANURE POLLUTION SOURCES

Sources that can contribute to water pollution problems resulting from improper manure management include runoff, leaching, poor land application techniques, open lots, and the improper location of manure storage pits or piles.

Manure application to your fields and pastures needs to be controlled in a way that does not allow the nutrients to seep into water systems. This can occur if manure is applied to frozen or snow-covered ground, saturated soils, sloping fields, areas that are too near to streams or ditches, or by using an excessive application rate.

Composting manure is an environmentally sound method of handling stored solid materials. Placing the compost on solid surfaces such as a cement slab or pit minimizes runoff and leeching before it is spread. Having the flexibility to store manure during periods you cannot get into fields assists in planning a spreading schedule to avoid times of high runoff from rain or snow.

Compost piles can be placed in low traffic areas and used as the need arises. Periodically stirring the pile will help transform the manure into more stable nutrients. Compost can be sold as mulch or garden manures to create another possible revenue source for your farm.

A holding area for semi-solid manure can be constructed near the area used for finishing pigs to provide an excellent way to capture the manure and prevent runoff until it can be spread on the fields.

Runoff of manure content can occur when rain or snow mixes with the manure on the ground or in an open lot where pigs are fed or congregate. Heavily used areas generally have little or no grass coverage to filter out these materials before they drain to a water source. Runoff into streams is more likely when the field selected for spreading the manure is located next to a stream, pond, lake, or other surface water rather than the manure and water source being separated by another field, pasture, or grass buffer strip.

Leaching can be a problem when the application of manure on a field or pasture becomes too concentrated and the nutrients enter groundwater sources through the soil structure. Your soil condition and type will largely determine the amount of nutrients that can be adequately filtered before it reaches the groundwater supply. Lighter, sandy soils will not filter as many nutrients as heavier loam and clay soils.

Open lot areas where pigs congregate in large numbers to eat, drink, or lie down have the greatest potential for runoff. Typically these areas have little plant growth to constrain water movement from a heavy rain or snow melt and are generally close to buildings where the volume of water from rain runoff can substantially add to the amount of water draining through the lot.

Manure storage facilities can also be a potential source of pollution, depending on their location. The bacteria, nitrates, and other substances found in stagnant manure have the potential for well contamination if they are located too close to the water supply.

While you may not be planning for a large number of pigs on your farm, it is useful to understand the problems that may occur and incorporate methods to prevent manure pollution regardless of the number of pigs involved. This is good herd management and is part of being a good steward of the environment.

MANURE HANDLING

There are three types of manure that can be produced on your farm: solid, semi-solid, and liquid. Each of these types needs to be handled in a different way to effectively use the manure.

A basic manure-retaining basin can be built so the solid waste is pushed out onto a cement platform for easy removal and to eliminate runoff.

Solid manure consists of a combination of fecal matter and urine that has been mixed with dry bedding materials, such as straw, hay, sawdust, wood shavings, corn fodder, or any other material that can become part of the bedding for pigs. Typically solid manure is handled with a manure spreader that applies the material as fertilizer on the croplands.

Semi-solid manure contains little or no bedding materials and has a consistency between liquid and solid, much like a thick pudding or slurry. This consistency creates some difficulty in storing manure for spreading onto the fields. One solution is to mix in bedding materials to get a firmer consistency or to use a sloped floor to allow the liquid to drain to a holding area and then store the solids in another area. By separating the parts, semi-solid manure becomes easier to handle.

Liquid manure contains no bedding materials and is typically a combination solely of feces and urine. This requires a storage facility, such as concrete lined pits or upright structures, that does not let any of the nutrients leach into the ground. In confined buildings there is a slotted floor system where the manure drops down into a pit below the building. This storage system requires pumping the manure from the pit into tanks that can be taken to fields and spread on the open ground or injected into the soil.

As pigs move about the pasture they deposit manure in many different areas. These manure clumps decompose and provide nutrients for plants and homes for insects that feed on the fibers and organic materials. The insects provide food for grassland birds to enhance nature's cycle.

Collection basins such as this can be used for containing semi-solid or liquid manure. This type of structure can also retain rainwater running through high-volume pig areas to prevent it from entering streams, roadside ditches, and farm valleys.

MANURE STORAGE OPTIONS

There are several options available for storing the manure produced on your farm. Larger farms with many pigs generally need larger storage capacities. With a smaller herd, storage options may be more affordable because the startup costs can be minimal.

If you are raising pigs in relatively small numbers, composting is one way to handle manure in an environmentally sound manner, and it can be a potential revenue source for your farm. Composting is the active microbial treatment of solid manure by using oxygen as the main catalyst. The organic matter is allowed to decay in a pile or windrow. Decaying organic matter creates heat, and a compost pile of manure, depending on its density, can reach temperatures of more than 160 degrees Fahrenheit at its core. Oxygen is required for composting, so the pile needs to be occasionally turned or stirred for the material at the edges of the pile or windrow to become part of the heating process. A tractor with a bucket loader, a skid-steer loader, or other equipment designed for stirring piles or windrows can accomplish this task.

Your pigs can also be used to stir the compost piles by placing small amounts of corn in holes burrowed into the row to encourage the pigs to root through the compost. As they stir the pile, it can be reshaped after you remove the pigs from the compost area.

Composting has the advantage of reducing the volume of manure and transforming it into a more stable nutrient form. These nutrients, when spread on the fields, are slowly released into the soil for crop nourishment. Because of this the nutrients in composted manure are less likely to be transported off the site through runoff and leaching into the ground water.

Another advantage of composting is that manure can be stored until weather and field conditions are better for hauling and spreading and the soil has a greater absorption rate of the nutrients. Composted manure can also be sold as an off-farm fertilizer, soil additive, or mulch. Garden stores offer these products in plastic bags for gardeners and vegetable growers. Because the manure has been broken down into less volatile nutrients, it provides a more benign product to be used by the general public. Developing a market for composted manure may relate more to distribution than actual production. Information regarding this market and how to develop it is generally available from your county agricultural extension office.

Fabricated steel manure storage structures can be constructed to hold large volumes of liquid manure produced from confinement programs. Storage sizes are generally dictated by regulatory agencies in most states and are usually large enough to hold at least six months of accumulation in many of the Midwestern states. This reduces or eliminates the need to spread manure during unsuitable weather and climate conditions, such as when the ground is frozen or when fields are too wet to support heavy machinery.

Knife attachments can be added to the rear of a tractor-drawn liquid manure spreader. These create lateral furrows below the soil surface as the machine passes over the field. Liquid manure is deposited in a large underground area and minimizes any manure runoff since it is buried beneath the surface.

Rolling or steep hills need a crop rotation plan that can protect soil by minimizing erosion while producing sufficient amounts of feed for your pigs. Your county Farm Service Agency can provide sound advice about setting up your farm rotation plan.

PASTURE MANURE MANAGEMENT

Pastures are often overlooked as a source of manure runoff. The most economical way of spreading the majority of the manure produced by your pigs is to let them distribute it themselves in the pastures while they move about.

Pigs will spread their manure more or less uniformly around the pasture as they move around. This requires no extra handling on your part and saves labor and fuel costs of transporting the manure to be spread on the field. By rotating pastures, avoiding steep slopes, streams, and drainage ways, you can minimize any possible runoff effects. The manure pigs create will slowly decompose in the pasture and provide a habitat for a wide range of useful insects that help break down the manure into nutrients used by the soil.

Buffer strips along stream banks adjacent to fields with a potential for runoff can reduce or eliminate the amount of manure entering a stream. These grassy strips help stop, filter, and hold back the sediments of runoff. Fencing plays an important role in buffer strips because it keeps pigs from entering the banks and waterways.

MANURE MANAGEMENT PRACTICES

Manure or nutrient management practices are generally required for farms expanding beyond a certain threshold for pig numbers or those involved with government farm programs. Nutrient management plans are developed in consultation with experienced professionals and must meet government approval. These plans develop a program for each farm where the nutrients produced from manure are accounted for in the total field application of fertilizers, whether they are purchased or come from the animals.

Accounting for nutrient application on all fields ensures that excessive amounts are not used and the possibility of leaching or runoff into the groundwater supply is limited. Areas of high vulnerability for runoff are identified, and spreading manure in those areas is restricted. By not exceeding crop requirements, the soil is not saturated with nutrients it cannot absorb. A good manure management plan includes soil tests, which are the starting points for determining the nutrient content of your fields. The results explain the nutrient requirements of each field and serve as a guide to manure application rates.

Assessing crop nutrients already available in fields from manure, legumes, and organic wastes is another component of good nutrient management. With a calculation of the amounts already present, the total available nutrients can be deducted from the soil test recommendations and will help determine what, if any, additional fertilizer purchases are needed.

SOIL CONSERVATION PLAN

If you plan to participate in any federal farm programs, a soil conservation plan is required. The conservation plan is an important part of any nutrient management program because it identifies crop rotations, the slopes of all fields, and the conservation measures you will need to follow to stay within the tolerable limits of soil erosion.

This plan also identifies what fields may have restrictions for spreading manure because of proximity to waterways, especially in the winter. One component of this plan includes identifying the best time of year to spread manure, which will depend upon the manure-handling system on your farm. A farm with manure storage has a different plan than one that requires daily manure hauling.

This should not be a concern if you raise a small number of pigs. Many counties require a nutrient plan for any farm constructing new manure storage facilities or expanding its swine operation.

PLAN AHEAD

Many years ago the United States pork industry recognized that environmental issues would be critical to the survival of pork production within our country. The National Pork Producers Council (NPPC) became proactive and developed an education-certification program known as Environmental Assurance to provide a basis for informed decisions. Information from the NPPC may provide answers to questions you have about environmental issues surrounding pork production on your farm.

Many problems with farm manure can be avoided with a plan that involves the best use of manure over your entire farm, while staying away from potential runoff into waterways. Nutrient and manure management plans and a conservation plan for your farm can be developed with help from your county agricultural extension agent or the Natural Resources Conservation Service (NRCS).

Water is one of our most precious natural resources. It is everyone's responsibility to help keep it clean. Providing buffer areas between waterways and pastures is an excellent way to prevent nutrient runoff and soil erosion.

We are the custodians of our natural resources and must use them wisely and treat them with respect. We have benefited from the stewardship of those who came before us, and we are obligated to do the same for those who follow.

CHAPTER 5
BREEDING AND REPRODUCTION

Depending on the type of production system you decide to use—farrow-to-finish, farrow-to-wean, or wean-to-finish—reproduction will pay a significant role in all but the last system. Reproduction is a direct concern if you choose farrow-to-finish or farrow-to-wean production. When you buy feeder pigs from another producer to grow to market weight, your concerns about their reproductive capacity are probably limited. If you are raising sows to produce two litters each year, you have a vested interest in the reproductive performance of your sow herd.

The performance of your breeding herd, regardless of size, is fundamental to the financial success of any swine operation. By achieving high reproductive performance, you will be able to maintain your breeding herd, encounter fewer problems resulting from infertile females, and have more pigs to raise or sell. Understanding natural reproductive processes will enable you to use those animals you have chosen, regardless of breed, to repopulate your herd even after selling some of them over the course of the year.

REPRODUCTIVE ANATOMY

Pigs have a similar reproductive system to that of other mammals. The position and utility of the reproductive organs serve the same functions. Understanding the positioning of the reproductive organs will help when breeding your sows with a boar or by artificial insemination. Reference books with complete reproductive diagrams are readily available and can be consulted for further study.

At farrowing time, the sow's mammary glands swell and fill with milk. At this point good nutrition is very important and plenty of clean, fresh water should be available to her at all times.

ESTRUS

The female reproductive system of the pig has two ovaries that lie just in front of the pelvic cavity and are similar to a kidney bean in size and shape. The ovaries are intimately related, work together, and are attached to the tissues that envelop most of the other reproductive organs. When the sow is open (not pregnant) her ovaries lie within or on the front edge of the pelvic cavity. In advanced pregnancy, the ovaries are carried forward and downward into the abdominal cavity along with the enlarged uterus that contains the fetuses.

The ovaries, through the production of ripe ova, or eggs, set the pace for the rest of the reproductive tract, which must be adjusted to receive the fertilized eggs and carry the developing fetuses through to birth.

The ovaries have a dual function: the production of eggs and the secretion of hormones that cause necessary adjustments in the other parts of the reproductive tract to take place. The heat cycle, or rhythm, is divided into several more or less well-marked phases.

The first phase is the development period when the pituitary located near the base of the pig's brain secretes follicle-stimulating hormone (FSH) and luteinizing hormone (LH) to initiate and stimulate the ovary's activity to start the process. FSH causes many small follicles, called oocytes, to grow within the ovary, principally due to an increase in fluid. The oocytes appear on the surface of the ovary and resemble a water blister.

Each follicle contains an egg, and as it develops, considerable amounts of hormones are produced, most notably estrogen. The follicles continue to grow, and as the estrogen levels increase, the eggs begin to mature. The elevated estrogen levels in the follicle lead to increased estrogen levels in the blood. When the estrogen concentrations in the blood become high enough, the female shows signs of estrus. The estrus or heat stage follows. This is the period of desire and acceptance of the male and is well-marked in pigs by the standing posture by the sow or gilt.

The follicle on the ovary ruptures after the heat period has passed. The average time of ovulation is between twenty-four and thirty-six hours after the start of standing heat for gilts and thirty-six to forty-eight hours for sows. After the follicle has ruptured, the egg travels down the oviduct where it is fertilized with sperm, midway through. The former follicle cavity forms a corpus luteum that will produce and secrete progesterone to prevent further secretion of FSH and LH by desensitizing the anterior pituitary to gonadotropin-releasing

Swine Gestation Table (Based on 114-day period)			
Date Bred	Date Due	Date Bred	Date Due
Jan. 1	April 25	July 1	Oct. 23
Jan. 15	May 9	July 15	Nov. 6
Feb. 1	May 26	Aug. 1	Nov. 23
Feb. 15	June 9	Aug. 15	Dec. 7
March 1	June 23	Sept. 1	Dec. 24
March 15	July 7	Sept. 15	Jan. 7
April 1	July 24	Oct. 1	Jan. 23
April 15	Aug. 7	Oct. 15	Feb. 6
May 1	Aug. 23	Nov. 1	Feb. 23
May 15	Sept. 6	Nov. 15	March 9
June 1	Sept. 23	Dec. 1	March 25
June 15	Oct. 7	Dec. 15	April 8

Farrowing dates for other breeding dates can be easily interpolated. Producers should be ready for farrowing prior to the due date because of individual variation in gestation.

A breeding chart shows the date of breeding and the time to expect farrowing if the sow becomes pregnant from the day of service. It takes 114 days to fully complete the gestation period for pigs. The intervening dates can be interpolated from the dates given; for instance, a sow bred July 6 will farrow October 28.

hormone (GnRH), which stimulates the release of FSH. This action suppresses further heat periods from occurring and interfering with the follicle process already in progress. The high level of progesterone then causes the female to be uninterested in the boar.

The ovulation rate or number of large follicles that grow and are released at estrus is important because this number becomes the first limiting factor to litter size. While sows develop hundreds of small follicles during an estrus cycle, typically only ten to twenty follicles ovulate at estrus.

After fertilization occurs, the egg implants into the lining of the uterus and the fetuses develop. If fertilization fails to occur, the corpus luteum regresses in size and loses its influence on the process. This allows the GnRH levels to increase and stimulate the release of FSH as the whole cycle starts over again; approximately twenty-one days later the sow or gilt will again exhibit signs of estrus.

PREGNANCY RATES

After fertilization, the early embryos enter the uterus on the fourth day and remain free-floating and mix with each other until about day twelve or thirteen. Under normal conditions and with healthy sows and fertile boars, fertilization rates typically reach 95 percent. Therefore many embryos enter the uterus. However, not all embryos are equal. Some are defective and others are slow to develop.

The time when most embryos are lost is between days twelve and twenty of implantation. If a sufficient number of embryos survive, they secrete estrogen as a signal, which prevents the mother from releasing prostaglandin from the uterus. Prostaglandin is a hormone that will destroy the corpus luteum, suppress progesterone production, and result in termination of pregnancy. If all embryos are lost or too few survive to produce a signal, the sow will return to estrus at an irregular interval after mating.

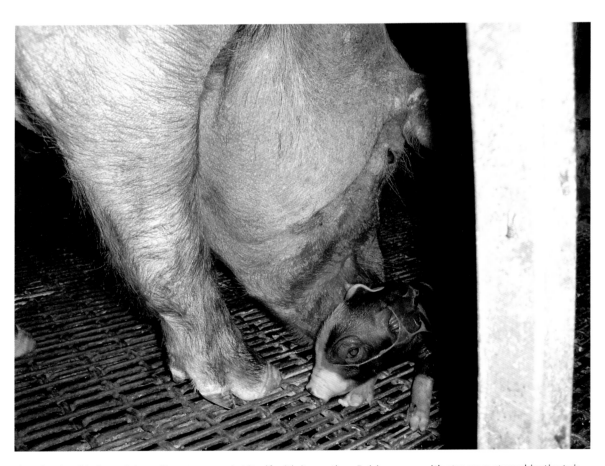

Shortly after birth, a piglet will try to acquaint itself with its mother. Raising sows with strong maternal instincts is an advantage to your herd.

```
                    Paternal Grandsire

        Sire _____

                    Paternal Granddam


Offspring ____

                    Maternal Grandsire

        Dam _____

                    Maternal Granddam
```

This standard ancestral chart is called a pedigree and shows three generations of animals. The top side of the pedigree is referred to as the paternal side and the father of the pigs is called the sire. The bottom side is referred to as the maternal side and the mother of the pigs is called the dam.

This is a typical three-generation pedigree, which is standard in all species of animals. Understanding this basic structure will help you in evaluating pedigrees of any pigs you buy privately or at auction.

If the embryos are successful in signaling the mother, progesterone will remain high and uterine contractions will be suppressed until birth. The embryos will begin to attach themselves to the lining of the uterus between days fourteen and seventeen. During this time and particularly between days seven and twenty-one, it is extremely important that all stresses on the sow, such as mixing with other pigs, moving to new areas, and changes in her nutrition, be avoided.

In some cases the sow's uterus will accommodate more embryos and fetuses than can be supported to birth. The fetuses that cannot be supported will be lost between days thirty and fifty. These are the basics of the reproductive process. For a deeper understanding of the entire process, reading materials are readily available.

HEAT DETECTION

Your ability to detect sows or gilts in heat and provide them access to a boar or semen through use of artificial insemination will determine whether they will become pregnant. Heat detection can be accomplished by allowing a boar to have access to the sows or gilts and let nature take its course. If you choose this method, the accuracy of your records to know when farrowing should occur will be severely hampered, unless you check for breeding activity several times each day.

It is fairly easy to recognize the signs of estrus in sows and gilts. Females normally assume a mating posture when they are in heat. They will assume a rigid stance or brace themselves and set their ears as if listening to something behind them. You can confirm this posture by pushing your hands against her back or loin region. Any bracing posture performed is a good indication that she will accept the boar. Checking for estrus in gilts is a little easier than in sows. If penned next to a boar, gilts may walk the fence, appearing to look for him. The appearance of a swollen vulva on a gilt is also a good indication of being in heat.

It is a good management practice to house all females to be bred in pens adjacent to a boar. The presence of the boar will stimulate the estrus activity of the sows and gilts and will aid in heat detection. Sows normally stay in heat from forty to sixty hours, while gilts will show estrus for twenty-four to twenty-eight hours.

Good recordkeeping is essential to a successful breeding program. By identifying which females are in heat on a particular day, you can more accurately determine their farrowing date than if left to chance. The females that fail to become pregnant will normally return to estrus at twenty-one days post-breeding. Marking the dates in a calendar will alert you to keep watch on those females previously bred. If no heat activity occurs, you can feel comfortable they have settled, or become pregnant.

As the piglets are able to move about, they will seek out the udder of their mother for milk. You may need to assist by moving them close to the udder and introducing their noses to one of the teats.

Begin heat detection of sows that have weaned a litter of pigs approximately three to five days post-weaning. Check them twice a day for signs of heat. Be aware that the onset and duration of estrus varies between farms and sows and that nutrition programs and stress factors can affect all of these processes.

REPRODUCTIVE PERFORMANCE

The process of reproduction is complex, involves highly specific biological functions, and can be greatly influenced by external environmental factors, including diet, temperature, disease, and housing. Your swine management program will have a major influence on your pigs' reproductive performance. The better you do your job, the more likely you'll have a better result. The sow and boar must be provided with an environment that enables them to express their reproductive potential to produce large, healthy litters. No two pig farms are the same, and the level of success you achieve will be largely determined by how you approach your program.

The reproductive performance of your sow herd can be measured in three ways: the number of litters produced per sow per year, the number of piglets produced per litter, and the survivability of those piglets. Anything that can be done to increase these three components will help you achieve greater success.

For sows that have already produced one litter, the manner in which they are fed between the time of farrowing to the time they are bred again will greatly influence the total number of litters produced, the number of piglets per litter, and piglet survivability. For gilts having their first litters, the environment in which they are born and raised has much to do with their development to reproductive maturity and will influence the size of their first litters. This requires attention to the entire lifetime of the female and all stages of development.

LACTATION

The future of your swine operation begins with lactation. Having assembled a sow or gilt herd, you must provide adequate nutrition for them to be healthy enough to be bred. Underfed or poorly nourished sows and gilts will have poor reproductive performance, since most of what they eat will be used for body maintenance.

After farrowing, it is extremely important to provide enough nutrients for body maintenance and milk production. Good milk production requires a full feeding program. It is difficult to feed a sow or gilt too much while she is lactating. During this time the ovarian follicles for the next pregnancy will be maturing during late lactation at a time when litter demand for milk production is greatest.

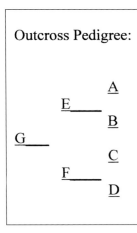

Outcross Pedigree:

E ___

    A

    B

G ___

F ___

    C

    D

This chart symbolizes that none of the six ancestors have any relationship with each other, assuming that those generations beyond this are unrelated. Outcross breeding refers to matings between unrelated pigs of the same breed. Matings between two different breeds is called crossbreeding.

To maintain the appetite of sows and gilts, the feed should be given in clean troughs, replenished frequently, and old or stale feed should be removed daily. Clean, fresh water should be plentiful, and if possible, control the temperature surrounding the sow. Temperatures too hot will depress food intake and cold temperatures will increase maintenance requirements. The metabolic rate of a lactating sow is high, and her core body temperature can rise significantly if the surrounding temperature is also high.

The body condition of the pregnant sow or gilt at the time of farrowing will have a significant effect on her performance as a mother. Excessive feed intake during the dry period prior to farrowing can depress lactation appetites. If sows or gilts are thin when farrowing, they will always be thin during their lactation because they cannot recover enough nutrients to gain weight. Striving for a balance between those two conditions will benefit the sow and litter during her lactation.

BETWEEN WEANING AND SERVICE

During the period between weaning, when all the piglets are removed from the mother, to when the sow is ready for breeding again, it is important to maintain feed intake to a level of about twice that of the maintenance level. The removal of her piglets will cause an involution of the sow's mammary glands as a signal for the start of the next estrus cycle. Good nutrition, particularly high fiber diets that include grasses and legumes (pastures), at this stage improves the maturity of the oocytes, the number of oocytes released, and the resulting early viability of the embryos produced. This ultimately increases the number of piglets produced and their

quality. There is a proven relationship between feed intake during the lactation period and the later average litter weight at birth. Having a good feeding program during this stage will help produce healthy litters and will help decrease the number of underweight piglets, or runts, at the next farrowing.

NUTRITION POST-SERVICE

After breeding, a sow can lose a significant amount of weight because of reduced feed intake. You should place weaned sows on a dry sow ration for a minimum of three days to keep them at a maintenance level. During their dry period, sows should be fed for body maintenance plus enough to add some condition to their bodies. This increase should be gradual and the feeding levels can be calculated with the help of feed company representatives. Your county agricultural extension agent can also help locate pertinent feeding information.

Nutrition for gilts having first litters is important. Since they do not have a dry period in the sense that sows with a previous farrowing do, gilts also need good nutritional preparation for breeding and their first farrowing. Good nutrition allows gilts to grow and mature at acceptable rates so they reach breeding age at target weights. Furthermore, gilts in good body condition will be ready to withstand a period of lactation. See page 87 for suggestions on how to figure out rations for growing and breeding gilts.

While it may be a challenge with a system where sows are on pastures or housed in loose groups, the benefits derived from supplying the appropriate nutrition levels during this period are great. Paying attention to the details can provide greater profits.

MANAGING THE BOAR

Boars are essential to breeding sows and gilts if you do not use artificial insemination (AI). They generally have greater influence than sows because each boar sires more pigs than are farrowed by any one sow or gilt to which he has been mated. Plus, each female receives one-half her genetic makeup from her sire.

The breeding performance of a boar can vary at different times of the year depending on the circumstances. The weather, number of sows or gilts he covers, his size and age, the nutrition level received, and other factors can affect a boar. Most of these factors are the result of management practices and can be minimized by the program you design for your farm. Boars that fail to show interest in females should be quickly moved out of the herd and another boar secured. Failure to breed your females during target dates will result in fewer litters produced, which will significantly reduce your potential income.

Another factor that may diminish the performance of the boar is overuse, especially of young boars. To avoid overusing any particular boar, determine how many boars should be kept based upon the number of females to be bred during any five-day period. To maintain acceptable fertility rates, you should not expect an active, mature boar to breed more than five females per week. Young boars should breed no more than two to three females per week.

For the best conception rates it is recommended that sows and gilts be bred twice during their heat period. This typically allows a sow or gilt to be bred approximately every twelve hours or until they are no longer receptive to the boar. Healthy, sound, mature boars can be expected to mate twice daily or ten matings a week. Young boars should not be used more than once daily with a maximum of five to six matings a week.

Proper care of boars includes providing adequate nutrition for maximum fertility without over-conditioning, controlling temperature stress to minimize the effects of extreme hot and cold

conditions, and handling them in a humane way. The manner in which boars are treated is crucial to their behavior. Boars should not be teased or abused and should be handled in a way that best reinforces their mating performance. Boars can be temperamental, and although you may prefer an aggressive trait relating to sexual activity, you do not want one that can become aggressive toward you or a family member.

With these guidelines you can expect reasonably good service from your boars. Depending on the number of females to be bred during a certain time period, you may need to increase the number of boars you keep on your farm. Generally the boar that provided one service to a female also provides the second. Depending on the condition, age, health, and interest of the mature boar it may be advisable to use a younger boar for second service, although this is not recommended for purebred breeding programs.

Because boars normally breed several females during each breeding season, they can be exposed to many diseases and become carriers of reproductive diseases. Proper herd management should include a vigorous vaccination program with a constant surveillance of sows, gilts, and boars. A discussion of disease preventative measures with your local veterinarian should be part of your boar management plan.

Proper management of the boar and good nutrition and health is essential for sound reproductive performance. Without a healthy, physically sound boar, the chance of impregnating sows is diminished.

Two examples of inbred Pedigrees:

Example A:

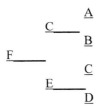

```
                    A
          C____
                    B
E____
                    A
          D____
                    B
```

This represents a full brother-full sister mating where A and B are the same sire and dam. This intensifies the relationship between ancestor and offspring by increasing the frequency of pairing of similar genes.

Example B:

```
                    A
          C____
                    B
F____
                    C
          E____
                    D
```

This inbred pedigree shows that sire C was mated back to his daughter E to produce offspring F. The inbreeding systems of half-sibling, full-sibling, and parent to offspring matings are the most severe types of inbreeding that can be practiced in livestock.

Inbreeding intensifies the relationship between ancestor and descendant by increasing the frequency of pairing of similar genes. These two charts show different examples of inbreeding. Above, full brother C and sister D were mated to produce offspring E. Below, sire C was bred to his own daughter E to produce offspring F.

WHEN TO FARROW

You will need to determine the time of year that is best for farrowing your sows and gilts. There are several factors that may influence your choice, including number of farrowings planned, number of females you have, facilities available, and production goals. It is possible for you to achieve two farrowings each year from your sows and mature gilts. From breeding to farrowing, 114 days is the average length of gestation.

Weather may be the most important factor for timing the farrowings on your farm. Generally it is best to avoid December to February and July to August as times to farrow your females unless you have climate-controlled facilities. These cold and hot months will require more attention to the sow and her piglets because of extreme temperatures. Cold temperatures require the use of heaters to keep the young pigs warm. Early in its life, the piglet's ability to withstand cold is limited. Piglet losses can occur quickly when temperatures remain below 60 degrees Fahrenheit at floor level. At temperatures below 35 degrees Fahrenheit, a fatal chill will occur within minutes unless warmth is provided. Drafts should be avoided at all times. Even with ideal temperatures, drafts can cause many problems for new piglets. Danger areas are wall cracks at or near floor level, open doors, and uncovered heat lamps in otherwise

The purpose of artificial insemination (AI) is to achieve sow pregnancies without maintaining a boar on the farm. However, some producers use AI in conjunction with a boar to try to achieve optimum pregnancies. AI technology has become more successful in the past decade and may be a consideration in your breeding program. The pipette used for AI is a hollow plastic tube with a sponge tip through which the liquid boar semen is passed into the sow's reproductive system. This tube is inserted through the vulva until positioned at the cervix opening.

cold buildings. These can create a draft at floor level when cold air displaces hot air. What may seem just a slight draft to you can be deadly for the piglet.

Similarly, hot summer months can cause severe stress to the piglets and sow. As the outside temperatures rise, the core temperatures of pigs also rise. Temperatures above 80 degrees Fahrenheit are considered undesirable for pigs. Farrowing sows and gilts in extreme hot weather or high humidity requires close observation to prevent heat stroke. Since pigs do not sweat, their ability to cool themselves is reduced. Adding the stress of

farrowing can cause complications to their body cooling systems. Some producers provide water sprinkling systems as a normal approach to keep pigs cool in summer heat. However, if the sprinklers are on at the time of farrowing, the cool water may not be good for the emerging piglets.

Avoiding the months of extreme temperatures for farrowings will diminish your risks and provide a better environment for the mothers and piglets. It will also reduce your workload by not having to make special accommodations for them. A well-thought-out plan will help eliminate stressful conditions.

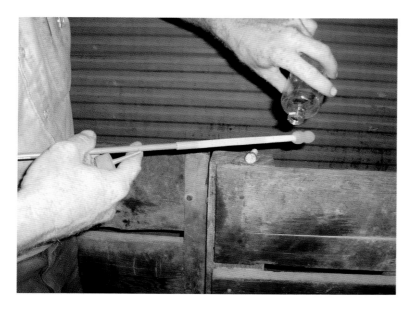

Lubricating liquid is spread over the tip of the insemination tube for easier insertion into the sow. Cleanliness is important and the tube should not be taken out of its protective packaging until you are ready to use it.

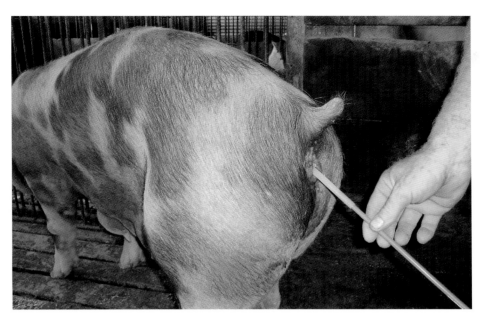

A swollen vulva is a good visual sign of estrus. The lubricated tip is inserted through it until it feels firmly positioned against the cervix.

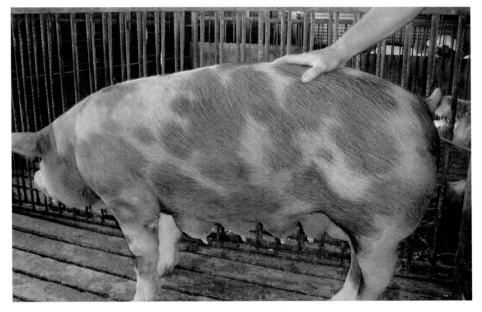

Using your hand to apply pressure against the back of a sow in heat will cause her to stand in a position as if she were accepting a boar. By pushing against her back, she will allow the artificial insemination procedure to easily and quickly be accomplished.

ARTIFICIAL INSEMINATION

In simplest terms, artificial insemination (AI) involves collecting semen from a boar and manually inseminating a sow to induce pregnancy. The technology has been developed to the point where conception rates and litter size show little difference when compared with natural service.

Swine AI has increased in use during the past decade. It allows superior boars to be used at a lower cost than some natural-service systems and with less risk of disease transmission. This requires less investment in a superior boar while reaping the benefits for a fee. One disadvantage of AI is that it may require a higher level of management than with your natural-service system. Because there is greater chance of human error, the conception rates may not be as high as with using a boar. Sanitation procedures must be strictly adhered to when using AI to prevent contamination of semen and equipment.

To utilize AI, you will need to purchase semen from a producer, probably located a great distance from your farm, to breed your females. Heat detection of the females must be done carefully and persistently to obtain satisfactory results. The semen is not frozen and stored in liquid nitrogen. It is shipped fresh and must be used within a short time for optimum sperm motility. The fresh semen is shipped in a bottle. After a plastic pipette or catheter has been inserted into the cervix, the semen is deposited into the cervix of the female via the pipette or catheter and the catheter is removed.

There are many steps involved with successfully using AI, but with training you can achieve high conception rates. More information is available from breed associations, boar studs, and your local county agricultural extension agent. With a conscientious effort to incorporate these practices, AI can work on any farm.

After the tube is properly inserted, the tip of the plastic bottle that contains the liquid boar semen is attached to the end of the hollow plastic tube and the semen is slowly squeezed through the tube into the sow's uterus.

As the semen passes through the tube and into the sow, it pushes a latex sleeve through the cervix to aid in depositing the semen into the uterus. A successful insemination will result in pregnancy, and the sow will not return into heat in 21 days. If unsuccessful, she will show heat three weeks after this insemination, at which time you will need to decide whether to use AI again or use a boar.

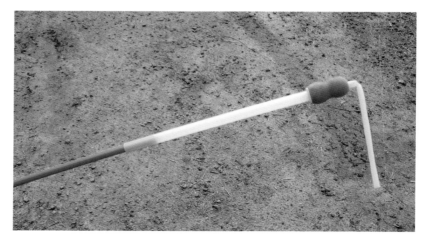

BREEDING SYSTEMS

There are several breeding systems you can use in your swine operation: inbreeding or linebreeding, outcrossing, and crossbreeding. Each has unique features, advantages, and disadvantages. Having a basic understanding of these systems may help you determine the method you wish to incorporate into your program.

Inbreeding mates animals that are closely related in order to increase the frequency of pairing similar genes. The physical expression of these genes will produce animals more similar in certain traits than the normal population. As an example, inbreeding involves the mating of a sire to daughter, dam to son, and littermates to each other—brother to sister.

Generally, inbreeding helps develop distinct family lines and is useful when crossing with other unrelated family lines. However, inbreeding tends to lower reproductive efficiency, survival rates, and growth rates. The reduction in performance is generally in proportion to the degree of inbreeding. Inbreeding also increases the frequency of undesirable traits or recessive genes more than normally seen within the general population. Inbred lines can be used to cross lines with no common ancestry and produce animals of superior quality, much like hybrids but without crossing with other breeds.

Linebreeding is a mild form of inbreeding and can be successful if used wisely. The objective of linebreeding is to keep a high degree of relationship between the animals in the herd and some outstanding ancestor or ancestors. In contrast to inbreeding, which generally does not attempt to increase the relationship between the offspring and any particular ancestor, linebreeding deliberately attempts to keep that relationship. An example of linebreeding is when a granddam is bred back to her grandson.

Outcross breeding is the use of any boar or sow from the same breed that does not have any relationship to any other pigs in the herd. Outcross-bred animals will exhibit the same general characteristics as the rest of the breed but have no close genetic relationship to each other. An example of this would be when two mated individuals have no common ancestor within the last five or six generations.

Crossbreeding involves mating two breeds to produce offspring that potentially carry the best traits of each breed. It can have a positive influence

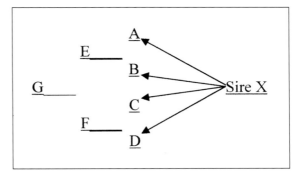

This chart shows one form of line breeding where sire X is the common sire of offspring A, B, C, and D. When these four are mated together, they produce offspring E and F, who when mated produce offspring G.

on economically important traits and reduce costs because of greater efficiency. An advantage that typically results from crossing two different breeds rather than using only one breed in a program is heterosis.

Heterosis, or hybrid vigor, is the measurable gain resulting from combining the genes of two different breeds over straight-bred parents. Generally the hybrid vigor effect can help improve traits, such as growth rate and fertility. Some swine breeders use two or three breeds and cross back and forth in each successive generation where others may use several breeds in a rotation. For example, a farmer may use Landrace and Hampshire for the first generation offspring and then breed all the resulting females to a Berkshire. The Berkshire-sired females are then bred back to Landrace, and so on. There are many different combinations and any successful crossbreeding producer will keep extensive records to track progress and identify trends in his or her breeding program.

CHAPTER 6
MANAGING AND SHOWING

As a pig owner you have the ethical responsibility to provide your herd with conditions where they can grow from quality feeds, reside without fear, and be treated in a gentle and humane manner. In some respects, the more you work with your pigs—in other words, the better you manage them—the more they may become an extension of the family. This may seem an odd consequence of raising them, but it is in your best interests, financially and philosophically, to have pigs that are content and easy to work with. Docile pigs will grow faster and be safer for you and your family to work with.

Rooting, when pigs tuck their heads low and use their snouts to dig up the soil, is a normal pig behavior. If handled properly, pigs can be gentle and display an animal affection for their owners.

Another aspect of managing pigs is learning to show them. Through showing animals, young men and women can learn a variety of skills not readily attained in other ways, along with becoming acquainted with other FFA and 4-H members from the county or state. These skills range from the daily care and handling of animals to time management, decision making, patience, and responsibility.

MANAGING PIGS

Good management requires using your skills and resources to provide your pigs with a safe, comfortable, and fulfilling existence. Pigs need day-to-day care and attention even though they may be self-sufficient in many ways. Daily care includes observing their physical state and condition and helping them avoid problems.

The best swine managers try to anticipate potential problems and take steps to eliminate them as quickly as possible. A small problem like a broken wire in a fence can quickly become a big problem—i.e., a hole large enough to allow a pig to escape—if it's not fixed right away. Although pigs are not as athletic as cattle, they can run fast. Trying to corral several pigs is a daunting task. Health concerns, such as illnesses, lameness, or other physical ailments, will need your prompt attention. Your ability to anticipate problems will come with experience and close observation.

In extreme weather you will need to observe your pigs several times during the day. Problems can surface quickly. Your response may determine whether a pig lives or dies and directly affects the profitability of your operation. Raising pigs is a business as well as a hobby, and as in most successful businesses, managers who pay attention to the minute details often have more success than those who don't.

UNDERSTANDING PIG BEHAVIOR

Pigs respond to the treatment they receive. Pigs that are comfortable around people have had positive experiences. Negative experiences with humans generally result in a large flight zone, which is the area where you can get close to the pig before it moves away. Positive experiences with humans will shrink this zone and allow you to touch, rub, scratch, brush, or easily handle them. Both experiences have advantages and disadvantages. If pigs run when approached, they may be difficult to handle when you move them from pen to pen, pasture to pasture, or lot to lot because of your inability to get close to them. Sows may not allow you near them when observation at farrowing may be necessary. One of the advantages of having pigs shy away from you, if it is an advantage, is that they most likely won't bother you when you enter their lot. On the other hand, pigs that are docile and tame may require more time during the loading, moving, or sorting process.

Providing sufficient space for the number of pigs you have in any pasture or pen will prevent crowding and aggressive behavior between pigs.

Pigs are gregarious by nature and can be quite social. They should be housed in groups of two to eight pigs if housed indoors or in larger numbers when housed outdoors. If your pigs need to be housed individually for any reason, it is wise to place them where they can see, hear, and touch other pigs through a fence or panel.

Young pigs will establish a dominance order with their litter mates, and this order will continue throughout their lives or as long as they are penned together. Pigs generally are mixed with other litters, resulting in times when this established order will be challenged by other pigs. This will result in fighting and threats until a new dominance hierarchy is determined. Once things are settled, subsequent encounters generally consist of only grunts and threatening postures. If it becomes a problem where one dominant pig refuses to let others eat, drink, or sleep, you may have to take action for the benefit of the entire group. You can identify the dominant pig by watching which one pushes at the shoulders and flanks of other pigs or makes them move away. Moving this pig to another group may lessen the stress on those left behind.

When serious fighting occurs, it involves teeth grinding, foaming at the mouth, and biting. This is more common in groups of uncastrated males and is good reason to separate them to different areas or pens. Adult boars may be very aggressive and should be housed alone, especially from younger boars. Separating two or more fighting boars is a dangerous situation because of the chance for injury from their teeth, not only for them but for their handler. Avoiding conditions where fights may occur is a good management practice.

Normal pig behavior is a rooting movement when pigs tuck their heads low and quickly raise it up and try to dig up or move whatever they are pushing with their snouts. Pigs are prone to use this motion when they are near their feed or water tubs, along the bottom edges of a pen or fence, or even with their handlers. Even if you have snipped off the incisor teeth at birth, pigs have strong jaws and the remaining teeth can inflict serious damage on another pig or you. Pigs can be curious, gentle, and display an animal affection if handled in a proper and humane manner.

MOVING AND SORTING TECHNIQUES
One key to handling your pigs is to understand that their eyes are placed well back on their head, which gives them wide-angle vision so they can see behind

Antibiotics must be properly stored and refrigerated when not being used. They have played a major role in treating pig illnesses for more than fifty years. While useful, antibiotics have become a mixed blessing as bacteria become resistant to them. Keeping a lock on the cabinet or refrigerator where you store the animals' medicines will prevent children from gaining access to the drugs.

themselves. They tend to refuse to enter dark places, such as doorways, unlit alleyways, or areas that throw shadows in their path. The use of lighting in the area you wish them to walk may alleviate their distress and make for easier movement.

One method for moving pigs involves using wood panels to block the way you don't want them to go. Placing a barrier in the path will make the pig seek an alternate direction, although it may try to lift the panel with its snout and seek an escape route.

Pigs typically do not have a herding instinct like cattle or sheep. If surprised or agitated, they may scatter in as many directions as their numbers so it is difficult to predict how the group will move. Often it is best to remove the dominant pig first, as that makes the rest of the group more submissive.

Make plans ahead of time for any move to diminish stress and excitement. If you are moving pigs from a large area, such as a pasture, to a smaller area, such as a holding pen, you may want to develop an alleyway where they can be sorted individually and then passed along to the smaller pen.

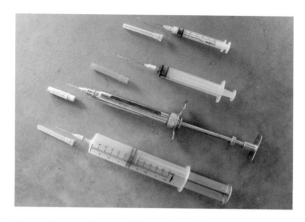

Depending on the specific need and dosage, injection syringes are available in several sizes. Needles should be discarded properly after each use to prevent the spread of body fluids between pigs.

Proper disposal of used needles is important to prevent accidental punctures and the possible spread of infections from one pig to another. Needles should never be reused.

It will be considerably easier to move your pigs if they are handled in a calm, relaxed manner. Some farmers use feed as an enticement to move pigs, although this may be more time consuming if the feed is not placed near the area where you want to move them. By placing feed on the ground several days before the move, you can get them to move toward that area by calling to them or banging the feed pail to get their attention. Making a repeated sound that they recognize and associate with food will make moving them easier than pushing or paneling them to an area.

COMMON PIG DISEASES

Pig diseases may be caused by bacteria, viruses, metabolic disorders, or parasites. There are roughly 140 different diseases that can affect pigs and a good manager will try to learn as much as possible about what may confront them on the daily rounds. The following is not a complete list of all pig diseases. Books that provide greater detail on these and other diseases affecting pigs, their symptoms, treatment, and recovery are readily available. A discussion with your local veterinarian or county agricultural extension agent may alert you to those diseases that might be of most concern in your area.

Atrophic rhinitis occurs in young pigs and the symptoms are seldom noticeable until about three weeks of age. There is no fever associated with the sneezing and discharge from the eyes, but the irritation of the nose can be acute as the pigs rub their snouts against their legs or sides of the pen. The progression of this disease causes a distortion of the face. The bones of the snout and face are often completely wasted away, which usually leads to a lateral curvature of the snout. A death rate of 20 to 30 percent is common for this disease. Treatment varies, but once an infection occurs, a cure is unlikely in that particular pig, although results vary when using sulfamethazine in the feed or sulfathiazole in the water for several weeks. Starting with a clean herd is the best prevention, and one way to prevent an outbreak is to test new pigs prior to bringing them home. A reliable test exists to identify carriers that may not exhibit outward signs of this disease.

Brucellosis, or Bang's disease, affects the reproductive system of the sow or boar. In sows it causes abortion and arthritis, and in boars it causes inflammation of the testicles. No fever is associated with brucellosis and death loss is very low in mature pigs, although the abortions cause the death of all fetuses. Brucellosis is spread by infections entering suckling pigs that drink milk from an infected mother. Water and feed can become contaminated from urine, infected afterbirths and manure, and discharges. If there is an infected pig in the herd, it may be impossible to keep the disease from spreading. There is no effective drug or treatment for infected animals other than removal from the herd. Developing a "Brucellosis-Free Herd," a term used in the eradication effort started by the United States Department of Agriculture (USDA) in 1961, is the most effective control of brucellosis.

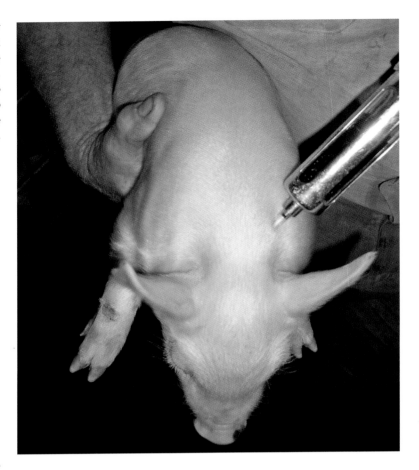

Acceptable injection sites for piglets include an area on the neck behind the ears and between the inner hind legs on young pigs. Always avoid inserting needles into the hams, loins, or shoulders to prevent puncture damage to these valuable meat cuts.

Hog cholera is a highly contagious swine disease that causes high fever, loss of appetite, and diarrhea, followed by coma and death in a high percentage of cases. Once the scourge of the pig industry, hog cholera has been successfully eliminated to the point that the last outbreak in the United States was in 1976. The disease is so important that, although it appears to be eradicated, any pig with a fever should be isolated because of the possibility that it could be African swine fever, an equally devastating disease that resembles hog cholera.

Parvovirus is widespread in the swine population, but the only outward sign of an infection is reproductive failure. It appears that the route of infection is oral, nasal, or perhaps venereal. Infection is generally associated with the embryos and fetal tissue where infection of the embryos at twelve to twenty days after conception results in embryonic death and a delayed return to heat. Other losses can occur during the different succeeding development stages with the end result being the same: loss of piglets and breeding time.

Vaccination is one method for controlling this disease, but maintaining a closed herd is the most desirable control method. There is no treatment for parvovirus infection, but it can be managed by exposing new boars and gilts to older sows that may be carriers at least two weeks prior to breeding. This will allow them to develop antibodies during a nonpregnancy time and become immune without endangering subsequent litters.

Pseudorabies, or mad itch, is a contagious disease caused by a herpes virus. The disease spreads by contact with nasal and oral secretions from carrier animals, in some instances feral pigs. It is lethal for piglets, which develop fever, paralysis, coma, and death within as little as twenty-four hours. In older pigs this consistently fatal condition is shown primarily by intense itching, sometimes to the point where the pig will literally rub or lick itself raw prior to death. Vaccination is available for use in states where the disease is endemic, but it will not prevent infection, only the clinical signs. Some states have developed pseudorabies-free status, which allows export to other states and countries. States free of the disease generally do not permit vaccination. Every pork producer should be aware that pseudorabies is a legally reportable disease. Other species, including cattle, goats, sheep, horses, cats, and dogs, can contract the virus from pigs, most commonly by bites. Infected animals die within forty-eight to seventy-two hours and often show symptoms similar to rabies, although the virus is not related to the rabies virus. These other species do not transmit the disease.

Porcine Reproductive and Respiratory Syndrome (PRRS) first became recognized in the United States in the mid-1980s and was classified as a virus in 1991. It can be confused with pseudorabies but doesn't exhibit the nervous symptoms. PRRS is sometimes referred to as blue ear disease because it affects the lungs and reduces the oxygen levels in the pig. It also affects the reproductive systems of infected pigs. Other symptoms include elevated temperatures, lack of appetite, late term abortions, coughing, and respiratory difficulties. There presently is no treatment for PRRS, although using antibiotics for three or four weeks can reduce or prevent secondary infections.

Swine erysipelas is principally a bacterial disease of young pigs between three to twelve months of age, but it can occur in older pigs. The bacteria live in the soil and once the infection surfaces on a farm, it is likely to appear again because the microbes live in urine and manure of infected pigs. Skin lesions typically appear on the neck, ears, shoulders, and belly as raised red areas in a definite diamond shape. The coloration can range from pink to red to dark purple. When the skin lesions appear, death may occur two to four days later. Low forms of this disease rarely cause death, but fevers can reach 108 degrees Fahrenheit, which may cause death from dehydration or secondary factors of high temperatures, such as organ failure. Erysipelas responds well to treatment with penicillin. Consultation with your veterinarian is recommended.

Swine dysentery, known by other names such as bloody diarrhea, black scours, and hemorrhagic enteritis, is often an acute and highly fatal inflammation of the large intestine of pigs. It is most prevalent in the Midwest and is usually traced to swine purchased from sales barns or to visitors from other infected areas. It is found mainly in young, growing pigs, although it can affect piglets and older pigs. It is spread by pigs ingesting feces from sick pigs or carriers. The first signs of infection show a decrease in appetite followed by a rise in body temperature. Diarrhea begins, rapidly becomes worse, and is frequently bloody. Losses in feeder pigs can range from 10 to 20 percent and pigs that recover tend to be stunted and unthrifty, which makes an economic impact. Control through medication has had varied success and is a continuing expense if used. Hygiene, sanitation, isolation, and having visitors wear disposable plastic boots are the best tools for reducing manure contamination and spread of the disease.

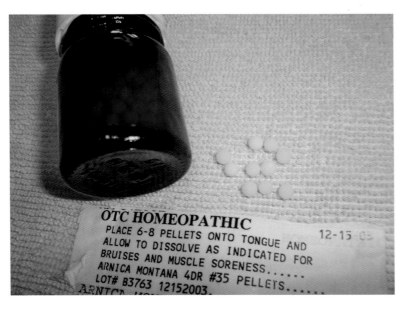

An alternative treatment protocol acceptable for organic production is homeopathy, which enhances the pig's immune system to fight off illnesses rather than suppress the symptoms. Homeopathic remedies are given as small pellets and work to elicit specific immune responses.

Iron is an important nutritional requirement for proper pig growth and utilization of feed. A piglet is born with a small amount of iron in its blood such that it must be introduced through other means, including injections that should be given within several days of birth. Be sure to read labels for directions and proper dosage. A multiple-dose syringe can be used for giving iron injections to a number of young pigs without having to refill. The dosage can be regulated by an adjustable marker on the syringe shaft. Be sure to use separate needles for each pig.

Transmissible gastroenteritis (TGE) is a major disease of pigs found most often in the Midwest swine-producing areas. It is a quick and debilitating disease caused by a virus that changes the intestinal lining, resulting in rapid fluid loss. It typically occurs in winter and early spring and can affect all ages of pigs, although its effect on piglets under two weeks of age can be devastating. The main symptoms include watery diarrhea and rapid dehydration, and its rapid spread through adjoining litters is a common sign. Death rates for young pigs can reach 100 percent but these rates are typically less if older pigs are affected. Older pigs shed the virus in their manure for several months after all signs disappear. Treatment has limited value because of the fast action of this disease and antibiotics are usually too late to arrest the impact of the virus. Some vaccines appear useful in combating this disease, but avoiding the introduction of the virus through new pigs and keeping unnecessary traffic near your farrowing areas to a minimum are the best prevention measures.

GOOD SANITATION AND DISEASE PREVENTION
Many health issues have the potential to affect the performance of your animals and even their lives. Good planning and management, along with use of common vaccines, will usually help avoid most disease problems.

Find a local veterinarian who includes pigs in their practice and consult with him or her about a herd health program. A list of veterinarians in your area or region is usually available from your county agricultural extension office. A veterinarian should be consulted prior to bringing any pigs onto your farm.

Steps you can take to prevent the incidence of disease in your pigs include buying healthy pigs and avoiding mixing hogs from multiple sources. Ensure that all required blood tests, such as pseudorabies, are negative before you buy them. You should receive a health certificate at the time of purchase that shows all tests and vaccinations have been completed. After purchase, make certain the pigs have been properly identified, and if you are having them delivered, be sure they are transported in a clean, disinfected truck. If you already have pigs on your farm and are bringing in new ones, it is wise to isolate newly purchased pigs for thirty to sixty days and at a distance of at least 300 feet from your other pigs. At this time the new pigs can be vaccinated if it has not already been done. Finally, keep visitors to a minimum or provide them with disposable plastic footwear before you allow them near your pigs.

Good sanitation safeguards help protect against disease, and attention should be directed to farrowing areas as a first line of defense. Disinfectants have little or no effect in areas that contain manure and dirt because the organic material in them will deactivate many of the ingredients. Farrowing pens should be cleaned and then disinfected between sows using the area. Before placing a sow in your farrowing pen, wash her with detergent and water, and pay particular attention to her underline. She may also harbor eggs and disease germs on her feet, body, and udder.

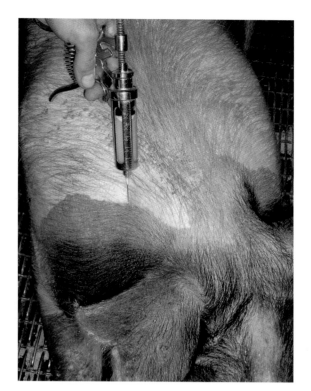

The easiest injection site for vaccinations or medicine on a sow or boar is behind the ears or in the neck. Unless restrained, they typically will try to snap at the perceived injustice and you should be careful not to get in their way.

Expecting sows and gilts should be thoroughly washed; otherwise the newborn pigs can swallow infective material with their first meal.

Isolating and treating sick pigs is one way of preventing an outbreak within your herd. Many disease outbreaks are the result of overcrowded conditions where pigs are raised continuously in the same facility that has not been thoroughly sanitized between groups, resulting in these organisms reinfecting the herd. Discharged fluids from sick pigs should be washed away immediately, and do not allow other pigs access to that area until it has been thoroughly cleaned and disinfected. If a sick pig has been held in a pen, the pen should be cleaned and disinfected once the pig has recovered or been removed.

Depending upon drugs to control swine diseases is a poor substitute for balanced rations, sanitation, and sound management aimed at disease prevention. Eliminating stress factors is also part of good swine husbandry.

Immunization Programs

Consider using effective vaccines in your management programs if they are available and you reside in an area with a prevalence of specific diseases. They may provide an effective disease-prevention program. Most vaccines are administered and distributed by licensed veterinarians. If you are planning to use vaccines, you should learn as much as you can about their effectiveness, dosage, and any other issues that might affect your particular situation. Consult with your veterinarian state swine extension specialist about the immunizations that might be most effective for your program.

PARASITE CONTROL

Internal and external parasites can cause problems for your pigs, as well as being economic thiefs. Pigs that are host to heavy infestations of parasites are more susceptible to diseases, such as scouring and pneumonia, which cause economic loss. The control of internal parasites, such as round worms, lung worms, and stomach worms can be controlled by using deworming agents available from commercial companies. Some products may have limited effectiveness, and there is no one product that is effective on all internal parasites.

External parasites, including lice, mange, ticks, and stable flies, can be controlled with various pesticide products if you have a conventional farming program. Organic protocols eliminate chemical pesticides, so an alternative treatment must be found. If you use pesticides on your farm, be sure to read all labels and precautions. Keep them away from children, pets, and other animals. Pesticides are poisons and need to be treated with care and respect.

HEALTH MANAGEMENT

One rule that can define any swine producer's experience is that pigs can get sick even with the best of care. The key is to minimize the number and severity of those illnesses to the greatest extent possible. This will start with inspecting your pigs every day. This observation does not necessarily require a great amount of time and can be done at feedings, when you will be able to quickly identify any pig that appears listless or exhibits other abnormal behaviors. Identifying such problems will allow you to move quickly to treat them. Generally a pig that exhibits symptoms of illness will not overcome it without your intervention or help. You become their diagnostician, doctor, and pharmacist all at once. While you may be able to

Record keeping is an important part of your pig-raising program. Whether you use a standard notebook or a more extensive computer system, records will help with making many important management decisions, such as breeding and which pigs to retain or cull from the herd.

Litter Number			Sow's Name			No.			Ear Mark of Sow

Recorded Litter No.

	Necessary Part of Pedigree of the Dam for Pedigree of this Litter

Owner of Sow When Bred — Dam's Name / Dam's Sire

Bought From Date — Far. / Number

Date Bred — Number Record / Dam's Dam

Boar — Breeder / Number

Date Due

Farrowed — Necessary Part of Pedigree of the Sire for Pedigree of this Litter

No. Pigs ____ Boars ____ Sows ____ — Sire's Name / Sire's Sire

Pigs Saved ____ Boars ____ Sows ____ — Far. / Number

Litter Ear Mark — Number Record / Sire's Dam

35 Day Wt. Date — Breeder / Number

35 Day Weight

PR Qualified: Yes ____ No ____

SLAUGHTER DATA ON PIGS FROM THIS LITTER

Ear Mark	Date	Wt.	Place	Length	Loin Eye	Fat Back	Qualified
							Yes / No
							Yes / No

RECORD OF PIGS SOLD FROM THIS LITTER

Ear Mark	Sex	Date	Sold To	Address	Price	Boar If Bred

Reg. No.'s of Recorded Pigs

Ear Mark	Sex	Number

NOTES ON THIS LITTER

perform some practices, a licensed veterinarian is required to administer some vaccinations, medical euthanasia, and certain antibiotics or steroids. Depending on your program, philosophies, and goals, there are three treatment systems that can be used on your pigs: conventional, homeopathic, and herbal. Each has advantages and disadvantages, and understanding the differences may help you decide which method to use in your program.

You should take care when administering medications, particularly by injection, because of marks that may be left on the carcass. Oral medications may be mixed with feed, syrups, or water. Subcutaneous injections are usually given in the loose skin where the neck and shoulder join. If intramuscular injections need to be given, they should be done in the neck muscles. Avoid injections into the hams, loin areas, and shoulders, if possible. Intravenous injections are usually done in the marginal ear vein.

Conventional Treatment

A conventional treatment program for pigs generally involves the use of antibiotics to relieve the symptoms of illness or disease. Although antibiotics may correct the problem in the short run, residual effects may affect marketing those treated animals. Thus, you should understand and know how to find the withdrawal time for the antibiotics you are using.

The use of antibiotics to treat disease in food-producing animals started in the mid-1940s, and by the early 1950s, the industry saw the introduction of antibiotics in commercial feed for pigs. In the past forty years antibiotics have served three purposes: as a therapy to treat an identified illness, as a prophylaxis to prevent illness in advance, and as a performance enhancement to increase feed conversion, growth rate, or yield.

Bacterial diseases generally cause pain and distress in the animal and are an economic loss if unchecked. Antibiotics can be used to help reduce

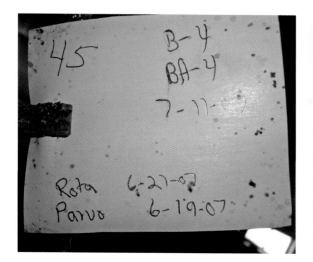

Placing cards above farrowing pens will help you keep pertinent information during the farrowing and weaning period. It can include the date of farrowing, number of live births or stillborn pigs, injection information, and castration dates. A simple system will help keep the information available until you can transfer it to your record-keeping system later.

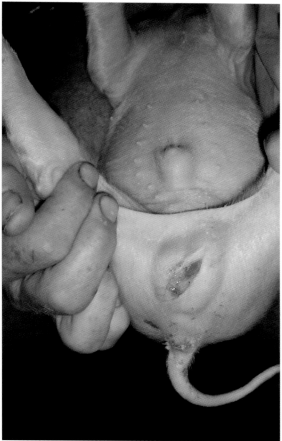

Castration is the removal of the testicles from the male pigs not needed for breeding purposes. An incision is made with a clean surgical knife and the testes are removed by pulling them from the body cavity until the cord breaks. This procedure should be done within several days of birth and extra care should be taken not to injure yourself, the person helping you, or the piglet with the sharp knife. Be sure to use iodine or a similar acceptable product on the cut to neutralize possible infections.

this suffering and distress and speed the recovery of an infected animal. When used responsibly, antibiotics can be an essential element in the fight against animal diseases. In rare instances antibiotics may be used to prevent diseases that might occur in a herd or group of animals if a high probability exists of most or all the animals becoming infected.

However, there are some cautions when using antibiotics, especially on a routine basis. In animals, as in humans, a significant proportion of those treated for infectious disease would recover without antibiotics. One study has shown that every year, 25 million pounds of antibiotics, roughly 70 percent of the total antibiotic production in the United States, are fed to chickens, pigs, and cows for nontherapeutic purposes, such as growth promotion. It is estimated that the swine industry alone adds 10 million pounds of antibiotics to its feed. This report also showed that the quantities of antibiotics used in animal agriculture dwarf those used in human medicine. Nontherapeutic livestock use accounts for eight times more antibiotics than human medicine, which uses about 3 million pounds per year.

The resistance of bacteria to drugs, such as antibiotics, has been documented in several studies. New antibiotics need to be manufactured as bacteria mutate and become resistant to the previous generation of drugs. The more antibiotics are used, the greater chances of a residue entering the food chain that can affect the general population. The excessive use of antibiotics can produce drug-resistant bacteria, which may become difficult or impossible to treat. Any pig that has received antibiotic treatment must be withheld from the market for a specified time and the withdrawal times must be followed by the producer or veterinarian.

This is not to say that antibiotics shouldn't or can't be used, but if they are used, it is imperative that they be used judiciously and only when warranted. In some specific cases it may be better for the pigs to be culled rather than put them through any treatment program, especially when it would likely be ineffective and uneconomical.

Familiarizing yourself with the diseases that require antibiotic assistance will help avoid unnecessary usage. A discussion with your local large animal veterinarian may provide other insights that will be useful in your program.

Alternative Treatments

Interest has been growing in using alternative treatment programs for pigs because of the concern of residual effects of antibiotics, both in pigs and the human food supply chain. Information is available for homeopathic treatment systems and this approach is ideal for those considering organic, sustainable, or biological farming methods.

Homeopathy and herbal treatments have a place within any farm health protocol. In years past these alternatives have been shoved aside for the quick fix of antibiotics that became inexpensive and readily available. The pressures of large-scale swine production and confinement have almost made it imperative to routinely use antibiotics to control any systemic problems. In some cases it is thought that using antibiotics routinely could replace good management practices. This does not have to be the approach on your farm. Homeopathy is based on the idea that bacteria are not necessarily bad and they do not need to be destroyed. Rather, it is not the illness that is being treated but the pig's reaction to the illness.

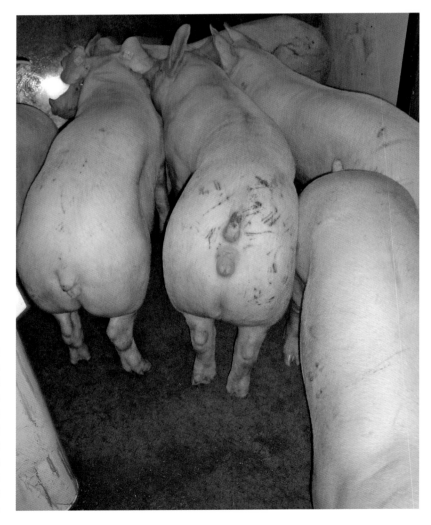

Many confinement systems dock or remove the tails within several days of birth. Crowded conditions are more conducive to biting behaviors and in these instances removing tails is a prudent management practice. Raising pigs outdoors tends to reduce this aggressive behavior and eliminates the need for tail docking.

Homeopathic treatment involves the natural stimulation of the pig's immune system so that it can fight off the bacteria that might otherwise cause a disease. It is known that antibiotics have a suppressing effect on the pig's immune system while it is fighting bacteria. However, it fights both the good and bad bacteria. Providing your pig with the ability to use its own body to fight off disease-causing bacteria benefits it in the long run because its physiological system is in better condition. Building up the health of the whole herd and increasing the resistance of its individual members to disease will help induce greater growth rates and meat production. A healthy pig produces healthy pork products. Two approaches can be taken when using a homoeopathic system on your farm: preventative and therapeutic (emergency treatment of individual cases).

A program of preventive medicine is better than treatment, and homeopathy is well suited to this approach. Homeopathy can be used to support the piglet in developing its immune system before it is even born by working with the pregnant mother. The first few days of a piglet's life are the most important and will determine, to a large extent, its health later in life. A sickly pig does not become a healthy sow overnight, but a healthy, vital piglet can have the ability to stay healthy.

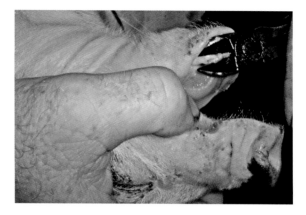

Clipping needle teeth is done to remove sharp teeth that may irritate the sow's teats or cause injury to other pigs as they play or fight.

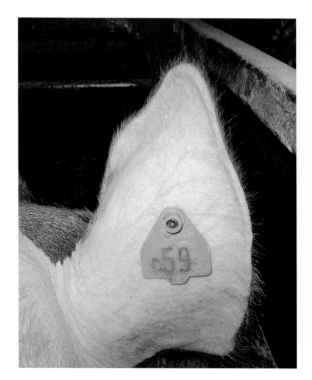

Records are only good if you can reliably identify each animal. There are several different ways to identify your pigs, including ear tags that are attached by using an ear tag applicator.

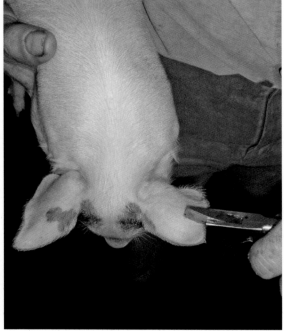

Ear notches have been routinely used for litter identification. Different notches on each ear can identify the litter and each individual within the litter. An ear-notcher is used for this identification method and should be done within a few days after birth.

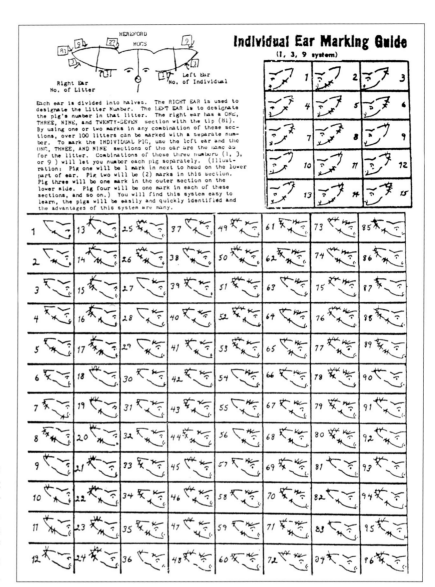

The position of the notch on each ear provides specific information for litter number and individual pig numbers. By studying and using this chart, you can make use of an ear-notching system for identification.

A homeopathic product is administered by means of a remedy. Remedies are derived from all-natural sources, including animal, mineral, or plant, and their preparation is made by a qualified veterinarian or pharmacist. A remedy is used in doses usually marketed in specific amounts and based on a system of potencies. In tincture form, the remedy is added to a sugar-granule base and allowed to soak, after which it becomes stable and can remain active for months or several years if properly stored.

These vials usually contain sufficient products for several doses, or administrations, for the animal. A dose may consist of five or six small pellets given at one time, depending on the condition being treated, and may have several applications during one day. The dose is placed directly on the pig's tongue and dissolved by the saliva. It does not need to swallow the granules because homeopathic remedies can be absorbed through the palate or tongue. The potencies are determined by the dilutions made from the refined crude product. This refinement develops the inherent properties of the specific remedy used for the specific needs of your pig. Homeopathy is a legitimate route for treatment of pigs, but is not a substitute for preventive measures like good nutrition, air quality, and proper sanitation.

A metal pig restraint can be a useful item around your farm and is easily moved from one location to another. Because they can be opened at both ends, you can have the pigs enter one way, work with them while they are restrained, and then release them at the opposite end to another pen or pasture.

A second alternative treatment is herbal, which also uses remedies based on preparations made from a single plant or a range of plants. Different application methods may be used, depending on the perceived cause of the disease.

These applications can be made from infusions, powders, pastes, and juices from fresh plant material. Topical applications can be used for skin conditions, powders can be rubbed into incisions, oral drenches are used to treat systemic conditions, and drops are used to treat eyes and ears. More information is available from alternative health stores or books published on these topics.

RECORDKEEPING

A good manager keeps a variety of records to help make business decisions, such as culling pigs or family lines that do not perform to expectations and pigs that are chronically sick. Managers keep identification records of individual pigs, breeding records to anticipate farrowings, vaccination records, and anything else that you consider useful. Paperwork is not necessarily a favorite job of many people, but keeping good records throughout your program will help you make intelligent decisions and identify potential problems. There are several computer programs available that simplify herd record keeping, or writing records by hand may be adequate for your needs.

Elevated, slatted platforms help maintain a clean area for weaning pigs to grow until they are moved to larger pens. These platforms can be lifted and moved with a loader to allow for easy cleaning.

Sunlight and fresh air are two of the most important ingredients for raising healthy pigs. Pens that allow them to be in sunlight and still protect them from the elements are an ideal housing arrangement.

Most people do not have a perfect memory and writing it down will avoid mistakes. As the size of your swine herd increases so, too, does the need for accurate records. You can use names, ear tag numbers, tattoos, ear notches, photos or any other common approach to identifying a particular pig in your records. Whatever makes sense and seems logical to you is most important, because these are your records to follow. It is useful, however, if the rest of your family can also follow the system, so keep it simple and keep it current.

If you have purchased pigs to begin your program, the best time to place some form of ear tag identification on them is when they have been loaded at the point of purchase or as they arrive at your farm. The few minutes that it takes to put tags into their ears will make their identification easier. Generally, one tag placed in either ear is sufficient for each pig. Although ear tags are commonly used, they get lost, the number fades, or tags become difficult to read because of mud, manure, or age.

Using photographs in making individual charts has the advantage of placing them at one point of time. As the pig grows and matures, having an early picture will provide a benchmark. Photographs may work as a form of temporary identification until you have time to apply ear tags.

Good management considers the comfort of pigs to be a primary importance. It can combine the use of heating lamps and heating pads to provide warmth for new piglets. A rubber mat placed under the sow provides a soft cushion to minimize the possibility of injury.

As each pig grows, its individual record can stay with it, and after several years you will have developed an extensive reproductive file on your herd. You will be able to track sows, how many of each sex they produce, and determine which sows appear to produce better offspring. This may factor into your considerations about retaining some sows and their offspring for future expansion because of desirable traits they exhibit. Other sows may not produce quality offspring you wish to keep. Your records can identify those you can afford to cull from the herd.

Vaccination records are helpful if there is a question about the health of any pig. They can safeguard the health of your herd, and the record of these vaccinations can help eliminate possible problems

A calendar can be used to calculate the approximate farrowing date for your sow. However, signs such as swollen mammary glands and a swollen vulva can alert you that farrowing may be only hours away. A good manager learns to identify these signs and keep a close watch during this time.

Skylights placed in a building with southern exposure can take advantage of the low winter sun, as the light can reach the back of the pen and provide warmth on cold days. This eliminates the need for an additional heat source and provides a natural environment to which pigs can quickly adapt.

SHOWING PIGS

FFA and 4-H programs across the country provide opportunities for youth to experience the pleasures and challenges of showing pigs in organized competition.

A pig project involves feeding, grooming, handling, training, and exhibiting the pig at a local, county, or state fair, or a breed-sponsored or sanctioned show. Depending on the quality of the animal or the interest level of the member, regional and national shows also provide excellent learning opportunities on a broader scale.

Prizes, premiums, and recognition are part of the benefits of showing, but the chance to work with an animal over an extended period of time has many residual, often intangible, benefits for the FFA or 4-H member. Showing pigs can help youth develop skills and learn lessons that can be useful throughout life.

MANAGERIAL OR OWNERSHIP PROJECTS

Two types of projects are generally available for FFA and 4-H members: managerial or ownership. A managerial project allows the member to raise, care for, and exhibit an animal without having to own it. This generally involves members whose parents, neighbors, or friends own the animal and allow the member to feed, train, and raise it for a specified period of time. This arrangement allows the member to learn how to properly care for an animal without the purchase cost. In this case, the member is generally responsible for all expenses during the length of the project, including feed, veterinary care, and insurance. In return the member receives all awards or financial premiums won at any show where the animal is exhibited. When the animal is sold to market, the owner receives a percentage of the sale price unless other arrangements between the owner and member have been made. An ownership project involves a member purchasing a prospect pig and assuming all costs during the length of the project and receiving all benefits including awards, honors, and total receipts at the final sale of the pig.

DEVELOPING LIFE SKILLS

Pig projects help FFA and 4-H members develop recordkeeping and budgeting skills because they assume total control of the pig. Sound records including all costs, expenses, income, and health of the pig are part of the record-keeping process to complete the project record at the end of the year. FFA and 4-H programs provide training and assistance to teach youth how to keep accurate records and serve as a model for real life experiences. Showing can be fun, but the lessons young men and women can learn from raising pigs are beyond financial measurement.

Pigs provide a unique venue for training when compared to other livestock. Dairy and beef cattle are halter-broke and tied-up during their stay at fairs and exhibitions; pigs are not. They are kept in pens and receive different training procedures that must be followed. This difference will involve discipline and patience by the member and the member's communication with the pig. The latter involves unspoken and verbal commands between the member and the pig and the use of a pipe or whip to signal nonverbal direction. All training procedures are intended to show a pig to its best advantage.

FFA and 4-H project members can participate in showing pigs at county and state fairs, and possibly regional or national shows. Participating in these shows provides youth with an opportunity to become acquainted with other project members from their county or state.

Working with pigs can help young men and women develop poise, leadership, responsibility, and many other skills that can be useful in future career choices.

Discipline involves the daily dedication to animal care and scheduling the time to provide proper nutrition, water, and housing to make sure the pig is comfortable and well fed. This ensures adequate growth so it is comparable in size to others within its own age group when shown. A pig cannot be expected to reason the same way as a person. Steady, consistent commands and routine handling will lead to less confusion on the pig's part when it is learning what is required for proper showing success.

PERSONAL GROWTH

Participating in shows allows the member to learn about the ethics of raising and showing animals in competition. Ethics are an important aspect of showing animals and help keep the competition fair for all exhibitors.

Rules and codes of ethics have been developed for the swine breed regarding practices that are allowed and those that are discouraged or not tolerated. Basically any human manipulation that alters the physical appearance of the animal, aside from the hair coat, is prohibited. The reasons for these rules should be obvious. The alteration of the animal's physical structure, whether it is the bone or muscle, is detrimental to the well-being and health of the animal and cannot be tolerated for the sake of winning a prize.

Learning the ethics of properly handling and raising animals is basic to the emotional and mental development of the member. Understanding the ethical choices available and learning to make them is one of the most valuable lessons young people

can learn from a 4-H or FFA project. Guidelines for understanding these ethical practices are available from breed associations, industry organizations, and county 4-H extension offices.

LEADERSHIP SKILLS

Showing pigs provides FFA and 4-H members with an opportunity to develop leadership skills. Clubs and chapters provide adult supervision to help members learn how to develop leadership skills, as well as their animal projects.

While under adult guidance, members have opportunities to lead discussions. In the case of older members, they can demonstrate to younger members the showing techniques and methods of raising animals that support the ethics training.

The importance of learning leadership skills, discipline, and patience at a basic level serves as a platform from which the member can build his or her personal integrity. This can have significant advantages when he or she later enters the job market. Having an understanding of animals and the ability to handle them correctly and ethically can provide members with many career choices not available to their contemporaries.

Careers in animal-related or scientific fields can be started with a project as simple as raising and showing pigs. People with experience in handling animals, ethics training, record keeping, and leadership skills are always in demand by employers. More information about the 4-H program is available from county 4-H extension offices. Information about the FFA program is available from a local high school chapter or your state FFA office.

CHAPTER 7
BUTCHERING AND MARKETING

Before entering into a pig-raising program, you should decide on your end goals. Do you want to raise pigs for your own use? Do you want to develop a niche market for your pork production? Do you want to raise pigs for enjoyment? Deciding early on what your intent is with the pigs will help you focus your efforts.

Developing a niche market for your products can potentially provide extra income for your farm. Products can be sold privately or publicly, at large or small farmer's markets, co-ops, or neighborhood groceries.

After pigs arrive on your farm, whether through purchase or birth, there are three routes by which they leave: selling them for slaughter, either for yourself or to a packing house; selling them to someone else; or death. Without a plan to selectively remove excess pigs, your farm will become overstocked by increasing pig numbers. By developing a marketing plan before you reach this point, you can use the increasing numbers of pigs for income, food, or both.

If you spend considerable time working with your pigs and raise them from birth to market weight, it is understandable that their final disposal may be a time of hesitation by you and your family. Can you say goodbye to them? While it may be momentarily difficult, the way you approach your business at the beginning can help turn these ambivalent feelings into a satisfying and positive experience.

Animals have always made a significant contribution to the welfare of human societies by providing food, shelter, fuel, fertilizer, clothing, and labor power. One of the unique aspects of pigs is that they are a renewable resource because they reproduce themselves for another generation and can utilize other renewable resources during their lives.

One defining ethic for pig owners is to be thankful for the lives of their animals and the purpose for which they were placed on their farm. This is an authentic discipline for the stewardship of any animal placed in your care.

If animals can experience fear and fright it would seem that they can also experience affection and love from their caretakers, something akin to human emotions. Although it is not believed they experience it in the same way or to the same extent as humans, it is our obligation to be thankful for their presence in our lives. Otherwise they become nothing more than disposable beings.

Responsible animal welfare is necessary for the long-term sustainability of the livestock industry. Consumers are more insistent upon and want assurance that animals are raised in a humane manner. Humane treatment shows pride in the care you give your pigs and that you are handling them in the best way possible.

The way pigs are handled has a direct impact on the quality and quantity of meat products they provide. Responsible care makes good economic sense, as well as reduces the levels of injury, bruising, and stress. The healthier the pig, the better the quality of the product, and the more profitable the pigs can be.

MARKETING OPTIONS

Marketing pigs includes all activities involved in the process of moving live animals and meat products from the producer to the consumer, including buying and selling. Marketing also includes the physical handling activities, such as grading, processing, and transportation. The term market has several meanings when applied to the swine industry. There is the hog market, which usually refers to the prices being paid at a particular place on a certain day. It may also refer to a market trend—higher, lower, or steady.

A hog market is also a physical place where hogs or feeder pigs are bought and sold. Hog Futures Market is when contracts are bought and sold for future delivery or purchase at the Chicago Mercantile Exchange and Mid-America Exchange, which is also in Chicago. The futures price is the price at which contracts are bought and sold.

Defining your objectives for marketing your animals that have reached a target weight will give you several options. Different marketing techniques have been used for many years, including feeder pig sales, local auction markets, and private treaty transactions between producers and order buyers.

Auctions have the advantage of establishing livestock prices through competitive bidding, although over the past decade these have been in decline with the introduction of electronic means like video auctions and the use of Internet markets.

You can use any of the above markets to enter into or exit from your swine operation. Either way, there are groups referred to as market classes that break down further into classes determined by sex, weight, and grade.

PLAN FOR YOUR MARKET

Choosing the time and method of marketing your animals can influence your profits. Too often producers sell pigs at the most convenient time rather than the most profitable time. To become a good pig marketer, you will benefit from understanding the marketing system and how the prices are determined, and then review the options available before you decide which to pursue. Keep in mind that it is the consumers that eventually have the most impact on market prices because of their tastes, attitudes, and sense of their families' welfare.

VALUE-ADDED PRODUCTS

There are several options available if you are unable to sell a large number of animals each year. A shift has been occurring from commodity production where small producers sell through auction markets and feedlot buyers, to direct selling a value-added product. While not a huge shift, this option may work for you, because by offering a value-added product, you can bypass traditional routes and capture more of the profits for yourself.

George's Fresh Kielbasa
Purebred Berkshire Pork
Antibiotic Free
Pasture Raised

Manufactured for: Willow Creek Farm, Loganville, WI 53943

In 1915, George Bar immigrated to Detroit from Poland. He brought an old family kielbasa recipe with him. Every Friday night he made his kielbasa to sell at his own grocery store. This recipe has been lost to the public since George's death in 1940. His granddaughter, Susan has resurrected the recipe using Purebred Berkshire Pork. The hogs are pasture raised on her farm using no antibiotics. Only the highest quality spices are used in this recipe. We hope you enjoy an old family tradition.

Cooking Suggestion: Cover links with water in a thick bottom pan. Bring to a boil, simmer for approximately 30 minutes. Let water cook down, turning links often so they don't stick. Add water, 1 tbsp at a time to brown links. Cut into 2-inch slices and serve with horseradish if desired.

INGREDIENTS: Pork, water, salt, sugar, spices, mustard, garlic.

KEEP REFRIGERATED

Recipes that have been passed down through families have played an important part in the past. As small-scale producers rediscover these recipes, consumers have more choices. This has enhanced the entire pork industry.

Selling pigs as a value-added product for a price over the market price means more profit for you. The best way to get a better price is to sell a better meat product. Consumers may be willing to pay more for a higher-quality product.

Value-added selling may require some marketing skills on your part. Find that niche market where you can sell your pork products. Finding your own customers generally requires more effort than traditional marketing systems, but the rewards can be greater.

Clearly explain the advantages of your hogs and your products to potential customers. Why are they better? It may be because you raise the pigs using a pasture-based program to enhance their growth and the eating quality and taste of the meat. You may use other concepts that capture the attention and imagination of potential customers, such as certified organic. Value is created when a product meets or exceeds the customer's expectations every time.

Niche markets for such things as natural, organic, or pasture-raised are becoming recognized by customers for the healthy conditions under which the animals are grown. Marketing groups may be available to help you sell your product in these markets. Most county agricultural extension offices can provide you with contact information for groups in your area.

The ham can be deboned for a desirable cut of meat, as shown here. Ham is a cut that can stand alone as a dish or be used in combination with other foods or for sandwiches.

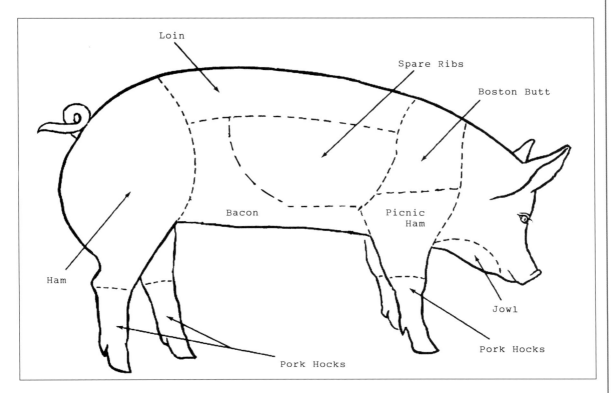

A variety of cuts can be derived from a pig carcass. The chart above indicates the areas on the body where the major wholesale cuts of pork are found.

BEST TIME TO SELL

The best time to sell your hogs is when they reach the target weight you have determined for your production system. There may be other factors to consider in timing the sale of your pigs, such as the time of year, market cycle, uniformity of the pigs to be sold, and your particular farm situation. The pork industry has established classes for marketing pigs and grades within those classes. This sorts the pigs into easily recognizable groupings to provide uniformity for buyers and sellers. While some grading applied to live pigs may be subjective, generally there is little variation within a given class or grade. The more experienced buyers and sellers can discern those minute differences. Having these uniform grades allows those who may not have that experience the confidence that they are getting what they buy. By understanding the grades and classes available, you will be able to decide the best time to buy or sell your hogs.

MARKET CLASSES

Market classes are typically set as feeder pigs and market hogs with additional breakdowns by sex, weight, and grade. Feeder pigs, either gilts or barrows, are normally grouped in 5- or 10-pound weight ranges beginning at 30 pounds and going up to 80 pounds. Pigs weighing under or over these weights are grouped and sold separately without regard to grade. There are two general factors for determining the grade of a feeder pig: an estimate of the grade the feeder pig will produce as a slaughter hog and whether it is healthy and thrifty. Feeder pig grades are listed as U.S. No. 1, No. 2, No. 3, and Utility.

Market slaughter hogs are classed the same as feeder pigs: by sex, weight, and grade. This class includes barrows, gilts, sows, and boars. Weight ranges are typically set at 230 to 280 pounds, which generally command the top prices, although this range can be set by the buyer and may vary from one location to another or from one plant to another. The top weight is based on the need for hams or loins. Pigs lighter or heavier than this range are usually discounted in price. Sows weighing 500 pounds or more usually command high prices because of their greater value for sausage.

Grades of market barrows and gilts are intended to directly relate to the grades of the carcasses they produce. Two factors are used for determining or evaluating the grade: quality of loin and the expected yield of the four lean cuts—ham, loin, picnic shoulder, and Boston butt. The lean cut yield is calculated as the weight of these cuts divided by the carcass weight multiplied by 100. Backfat thickness over the last rib and muscle thickness are used as a guide to estimate the grade. As an example, the standard Grade U.S. No. 1 carcass will yield 60 percent or more of the high priced cuts and have less than 1.00 inches of backfat. U.S. No. 2 will yield 57 to 60 percent and between 1.00 and 1.24 inches of backfat. U.S. No. 3s yield 54 to 57 percent and between 1.25 to 1.49 inches backfat, and carcasses graded U.S. No. 4 will yield less than 54 percent and have in excess of 1.50 inches of backfat. A carcass with an estimated last rib backfat thickness of 1.75 inches or over cannot be graded U.S. No. 3, even with thick muscle.

The market cycle can have an impact on your profits. If possible, try to avoid selling when many hogs are expected to reach market at the same time. In some cases it may be worth waiting a few days to sell your pigs if the market prices start to rise, as this may indicate a leveling off of numbers entering the packing houses. Familiarizing yourself with published hog statistics and market trends can help you identify the best potential market times during the year.

MARKETS FOR FEEDER PIGS AND MARKET HOGS

If you choose to sell your pigs when they are young, feeder pig markets are one channel to use. Typically these are graded auction sales where pigs are sold to other farmers or producers. Some hogs may be bought by order buyers or feeder pig dealers and transported to other markets. Generally pigs are placed in pens with other producers' pigs of a similar weight and grade. This commingling process allows larger groups of uniform weight and grade to be offered and producers usually receive higher prices for their pigs.

Private sales to other farmers are another way to market your feeder pigs. Some farmers prefer to buy directly from other producers because it lessens the chance for health problems and death losses than with commingled pigs from graded sales. With a limited number of pigs available as compared to large feeder pig auctions, selling only a few pigs at one time may present some marketing challenges. However, selling privately is a viable marketing tool and the prices received are usually based upon graded pig sale prices.

If you have developed a market for selling your meat locally and have devised a schedule to handle the number of pigs you produce each year, then you may not need the packing house as a vehicle for selling your pigs. Typically, excess pigs are sold to packing houses, sale barns, or a terminal market when they reach market weight.

Many pig producers sell their hogs directly to a meat packer and the prices received are usually slightly higher, because the buyer has fewer costs involved in acquiring the pigs, including transportation, death loss, and shrinkage. Packer buyers usually prefer to sort out any lightweight, heavyweight, or crippled pigs and separately price them. There are some markets that buy pigs based on grade and yield, which is derived from the carcass weight and grade. Producers using this system receive a higher price for better grading and higher yielding pigs. Conversely, the producers receive lower prices for pigs that do not grade or yield as well.

Sales barns can be used to market your pigs, but this option is used less by pork producers than other options. Pigs taken to an auction are generally sold by average weight and ownership group. Your pigs will typically be sold by themselves and sell on their own merit. Competitive bidding is one of the attractions that draw producers to use this channel. Pigs marketed through a weekly auction are typically purchased by packer buyers or order buyers for further shipment and sale to a packer and must go to slaughter because of health regulations to control swine diseases.

There may be costs, such as transportation, associated with where you send your pigs to market that are generally part of the seller's expense. One of the less visible expenses in transportation is called shrinkage, which is the loss of body weight in the marketing process. Shrinkage results from loss of excretions, such as feces and urine, which if held inside the pig's body would be included as part of the total weight. When expelled, it is no longer part of that weight. Shrinkage also occurs when a pig is away from feed and water or when it is subjected to a stress process, such as sorting, loading, and hauling. Typically the greater the distance and time in transit that pigs are hauled, the greater the percentage of shrinkage. Some of the excretory loss can be replaced with feed and water but the opportunity for such refilling does not typically occur at these markets.

Boneless pork chops have had the rib removed and are sold as a select cut. The pork chop is the most popular cut from the loin area.

Pork hocks are oval-shaped, two to three inches thick, and are cut from the picnic shoulder and are similar to pork shank cross cuts. They can be used for stews or can be braised or marinated.

The shoulder contains the Boston butt and picnic ham. The picnic ham includes the shank above the hock to the base of the Boston butt. The shoulder is generously marbled and is a fairly inexpensive cut.

The Boston butt can be cut to produce a pork blade steak or left intact to be used and cooked as a roast. The shoulder can be smoked if desired.

BUTCHERING AND MARKETING

The pork tenderloin is the most valuable cut of pork available. It is located on the sirloin end on the inside of the ribs. The tenderloin is versatile and can be cut into small pieces and sautéed or the whole tenderloin can be rolled and tied with string and roasted or braised.

The arm steak is thinly cut from the picnic ham and can include portions of the picnic roast if desired. These steaks can be grilled, broiled, or braised for meals.

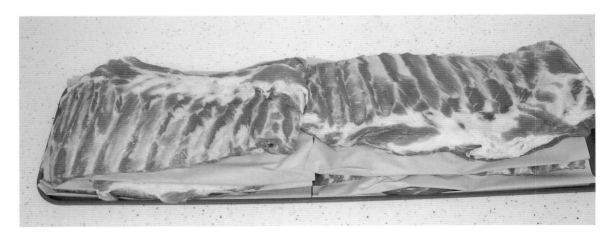

Spare ribs are removed from the belly and are the intact rib section after the primal cut is made. They can be smoked, grilled, or baked and basted with sauces if desired. After the spare ribs are removed, the rest of the pork belly is used for bacon. Slabs of spare ribs make a good choice for barbecue alternatives. They are a fairly inexpensive cut because the bone is included in the cost, and they are sold by the pound.

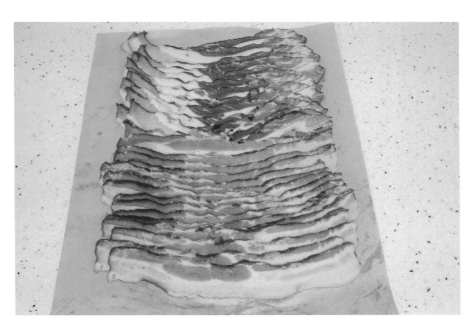

Bacon comes from one of the primal cuts—the pork belly. It may account for about 15 percent of the carcass weight, so it can be a substantial cut. The pork belly can be cured and smoked to produce bacon or it can be sliced fresh and frozen.

Packaging for your pork products can include strong plastic pouches that can be tied at the open end and filled with a predetermined weight of meat.

Freezer paper is a typical packaging material choice that can be used for single or multiple cuts wrapped together for consideration of serving sizes.

Ironically, pigs that have lost weight due to shrinkage are generally worth more per hundred pounds than pigs that have not shrunk. Producers will usually receive less for pigs that are full or have been filled prior to selling on a live basis because this becomes part of the shrinkage loss for the buyer.

Clear, plastic vacuum-sealed packaging is another efficient way to display and sell your meat products. They are easy to open and provide an excellent presentation of your cuts.

Small-scale pork producers can market their meat using labels developed to identify their farm or business. Developing your own label can be accomplished with the help of state meat inspection personnel or a local licensed meat processing facility.

MOVEMENT OF PIGS TO MARKET

Moving your animals to market in the easiest and fastest way possible will affect the quality of meat produced. Developing an area for loading hogs that moves them in a relaxed manner will help in defusing a fright reflex. Avoid using electric prods, clubs, sticks, whips, or anything that can inflict bruises when loading pigs. Bruised pigs have bruised meat, which needs to be trimmed during the slaughtering process, and the packer may attempt to overcome any visible bruise conditions by paying a lower price for the pig. If the sale is being made on a carcass weight and grade basis, the producer bears the loss of bruise-damaged or otherwise condemned carcasses. Feeder pigs are especially prone to bruising and rough handling, which can result in leg injuries.

Take care when hauling pigs for your own use or for sale to a customer, packer, or another producer. Some heavy-muscled pigs can have porcine stress syndrome (PSS) and die quickly if stressed in hot weather. If a pig appears stressed, allow it to lie down and rest. Never use an electric prod or other means of stimulation on a stressed pig or it could quickly die. Also, do not pour cold water on an overheated pig. This can cause shock and lead to death. In this case it is better to use a sprinkling can or spray nozzle on a hose to gradually cool the pig to the point of recovery.

You can haul your pigs to the slaughter site if you have a livestock trailer, or you can hire someone to transport the hogs. If you hire transport, make certain the hauler understands your concerns about animal movement and treatment at the slaughter plant. All of your work can be negated if the animal is mishandled or mistreated while being unloaded in your absence.

MARKET YOUR OWN LABEL

Capturing a niche market can have financial and personal rewards, especially if you develop your own meat product labels. There are labeling requirements, but they are not so excessive as to prevent you from entering this market. Private labels are becoming more prevalent as producers with small numbers of hogs find it to their advantage to use such production practices as outdoor or pasture programs as a selling point for their products. Since you may raise fewer hogs, you can provide more individual care and turn this into a solid marketing tool to identify your farming practices.

You will need access to a slaughter facility that can guarantee your pig's meat will be separated from others processed the same day. They can explain the costs involved in processing and packaging your meat with your label for proper identification. Be sure that the facility is state- or federally inspected.

After taking possession of the meat, you will need to locate or have adequate freezer storage facilities on your premises. Factors that influence the amount of storage space needed for the meat include the total pounds of meat from each pig and the number of pigs processed at one time.

Developing your label and storage facilities should be done well in advance of slaughtering any pigs, and your calculations should be made prior to purchasing or renting storage facilities so they are in place when your meat arrives. With your label and storage system completed, it will be in place to use whenever you have a pig slaughtered and processed for sale. If you plan to sell and deliver meat to customers who live a distance from your farm, you will require a storage container that can maintain frozen temperatures during the transport of meat to the customer.

There are popular meat products that can contain portions of different animal species and can be flavored with a variety of spices to suit many different tastes.

Some traditional meat products that have been popular with older generations are being rediscovered by those who are seeking more exotic table fare. A product such as blood sausage uses parts of the pig that many people don't often consider.

The concept of "locally grown" has become an attractive choice for many consumers who prefer to know where their food comes from. This has provided increased opportunities for small producers in the past decade.

Contact your state department of agriculture for details on how to start direct marketing meat. The agents there can advise you on all regulations and licensing requirements involved in selling meat under your own label.

ADVERTISING YOUR PRODUCT

Advertise your product when you have meat to sell. How you approach this, whether by word of mouth, print ads, or some other means, will determine how long your meat stays in storage. If you have more pigs growing on your farm it would be to your advantage to quickly move the meat in storage and have room available for the next round of carcasses. Help is available with your marketing and advertising program if you don't feel that you can handle it alone, but be sure to highlight the emphasis you place on a certain production system, which is appreciated by consumers.

SAYING GOODBYE

The final disposition of your pigs may cause some ambivalent feelings for you and your family. Raising hogs that you work with on a daily basis can be an emotionally rewarding experience for everyone. There is a certain type of communication that can exist between animals and humans, and to be part of that is an exceptional encounter with the world on a completely different level. Like humans, these animals develop their own personalities, and by extension, project it to those who work with them. There may be pigs that instinctively know what you want them to do and you may have pigs that walk across a pasture or lot just to receive attention.

It can be difficult to let your pigs go and turn them into food, but this is their cycle of life and you are part of it. In essence, you are helping them fulfill their destiny and one that was done with concern, care, love, and a sense of pride in providing an atmosphere in which they lived comfortable, healthy, and quality lives. There is no shame in that, and you can find great satisfaction in achieving these ends and by honoring their presence.

Whether you produce pork for your family's consumption or to sell under your own label, you will need appropriate equipment to keep meat frozen until it is used or sold. A chest freezer, pictured here, or an upright freezer can provide long-term service.

Transporting meat from your home to a customer will require containers that will keep temperatures below freezing. Chilling the insides of a carrier and including ice packs will help maintain the required temperatures.

HOME BUTCHERING

If you raise pigs to produce meat for your table, butchering at home may be a consideration for you. Butchering animals at home may be attractive if you value self-sufficiency, are adept at handling sharp knives, and have the necessary equipment, time, and expertise to complete such a project in a satisfactory and safe manner.

There are advantages to home butchering, but it is not for the faint of heart. It will take physical stamina because once the process is started it must be completed with little delay to prevent the fresh pork from spoiling. It will take equipment specially designed to process an animal and an understanding that the size of this equipment will be directly related to the size of the pig being butchered. It will also take some fortitude on your part to be able to quickly and humanely dispatch the pig with care taken not to injure yourself or others in the process.

The basics of home butchering are described here but with the understanding that this is not an endorsement of this procedure. Anyone contemplating using this process on his or her farm should take considerable time to completely understand the necessary procedures required, the parts and cuts of meat that can be harvested from a pig, the storage facilities and space required to preserve the meat, and whether you are physically able to handle the weight of the meat involved. An alternative is to locate a local licensed meat locker that can offer a professional butchering service.

For example, if the pig's live weight is 240 pounds and it has a 75 percent dressing weight (the amount of meat and bone left over once eviscerated), the carcass weight will equal about 180 pounds, or 90 pounds per side, since the carcass is cut down the middle. However, you will have to handle the entire 240 pounds once the pig is killed to begin the process. Having several people involved with the entire process will help with the work required for lifting.

A 240-pound pig, then, will provide approximately 108 pounds of meat for use by your family. The 108 pounds represents a 60-percent cutout from the carcass weight, because even after dressing out, there will be about 72 pounds of wastage—parts that cannot readily be used, including fat trim, bones, and skin.

The largest part of the carcass is usually the ham, which can be about 23 percent of the carcass, or in this case, 24 pounds. The side or belly and the loin areas represent about 15 percent each or 33 pounds; the picnic shoulder and Boston butt are each about 10 percent or 22 pounds; and the miscellaneous portions, including jowl, feet, bones, skin, and fat, account for about 25 percent of the carcass weight, or 27 pounds. While there may be some variance between pigs, these percentages generally hold true for normal, well-developed pigs of that weight range.

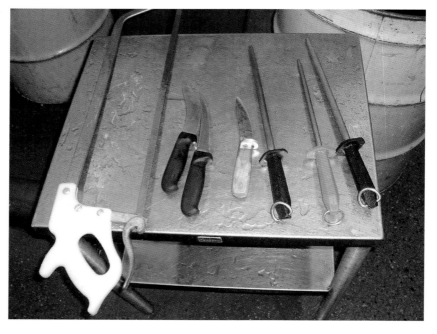

Clean, sanitized knives and saws that will be used in the butchering process should be laid out on a clean and sanitary table prior to starting the process. A meat saw, skinning knives, a sticking knife, and sharpening steels should be kept in an easily accessible area.

CUTS OF MEAT

The five major areas where cuts are derived can be further broken down into retail cuts often seen in markets. The picnic shoulder includes the upper front leg above the knee. This cut lies just below the Boston butt and contains a higher level of fat than other cuts, but it makes a flavorful and tender portion. The picnic shoulder can be smoked and cured to make the picnic ham, which is then ready to eat cold or hot. The arm and shank bones make up the shoulder and create a high ratio of bone to lean meat. When well-trimmed, this cut is used for lean ground pork and can be cubed or cut into strips to use for kabobs, stir-fry, or stews.

The Boston butt, also called the shoulder butt, is often a better cut than the picnic shoulder. It lies at the upper portion of the shoulder from the top to the plate to make the backbone. This cut is tender, full of flavor, and can be cut into roasts with the bone intact or cut out for boneless roasts. The roasts can be cut into blade steaks that can be broiled, grilled, or braised.

The pork loin cuts are located directly behind the Boston butt and include a portion of the shoulder blade bone. The loin includes most of the ribs and backbone all the way to the hipbone at the rear. There is a loin area on each side of the pig, and together they will account for about 20 percent of the carcass weight. Many retail cuts are derived from the loin, including top loin roasts, pork chops, baby back ribs, pork tenderloins, loin chops, rib chops, and blade chops. The loin constitutes a long strip that contains the top section of the ribs. When these are trimmed away, the result is a boneless pork loin. The section of the loin between the blade end and sirloin end is usually referred to as the center, as in center chops and center roasts. A boneless pork loin is smoked to produce Canadian bacon. Rib bones trimmed from the loin can be barbecued as pork back ribs. This is also the area where pork backfat is located. This is the thick layer of fat between the skin and the eye muscle, which may have some cooking uses but is used mostly to help determine carcass grades.

Hams, which make up about one-quarter of the carcass weight, come from the rear leg area of the pig and include the aitch, leg, and hind shank bones. This is a prime cut area of the pig because it contains little connective tissue. Hams can be deboned and the shank portion of the ham, called the ham hock, is used the same as the shoulder hock.

The pork belly, or sides, are where the bacon and spareribs are cut out. They are located below the loin on each side of the carcass and account for about 15 to 20 percent of the carcass weight. It contains a lot of fat with streaks of lean meat. This area provides the spareribs, which are separated from the rest of the belly before cooking.

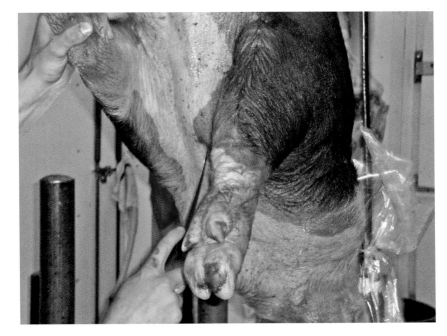

A clean kill begins by positioning the sticking knife in front of the point of the breastbone and slicing the jugular vein in the neck with one swift vertical motion.

The feet can be removed by using either a meat saw or an extremely sharp knife. It is advisable to wear a chain mail glove on the hand that isn't holding the knife.

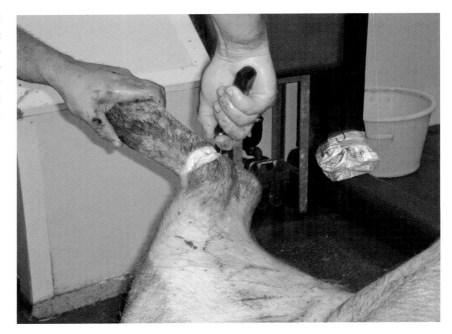

The skinning starts in the area between the front legs at the breastbone. By slicing upward and away from the carcass, you minimize the chance of puncturing the body cavity.

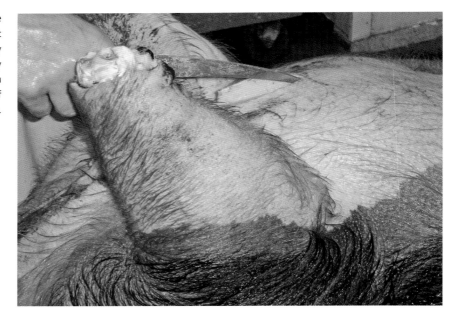

The miscellaneous portions of the carcass include the jowl, pig's feet, tail, neck bones, skin, and fat. Some cooks highly prize these areas, but the parts are often dismissed by the general public as unusable.

Eating raw pork is strongly discouraged because of the presence of the parasite *Trichinella spiralis,* which causes trichinosis in humans. These little roundworms migrate into pig muscles and mature into an infective stage. Unless destroyed by minimum cooking temperatures of 140 degrees Fahrenheit, they will be viable parasites for infections. Most experts recommend a minimum of 150 degrees Fahrenheit for cooking pork because of the inaccuracy of many thermometers. Trichinosis infections in humans can cause nausea, diarrhea, muscle pains, and aching joints. It is a treatable infection but nevertheless is an uncomfortable experience.

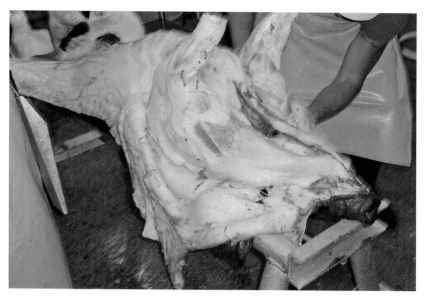

A skinning knife will slowly excise the skin from the body with minimal effort. While using your knife, pull the skin away from the body to provide the clearest separation.

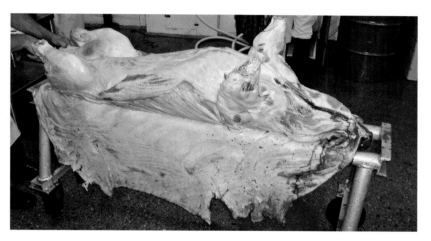

During the skinning process, you can leave much of the subcutaneous fat on the remaining carcass. This is more easily removed after the carcass has been properly cooled. Leave as much fat as possible on the carcass and not on the hide.

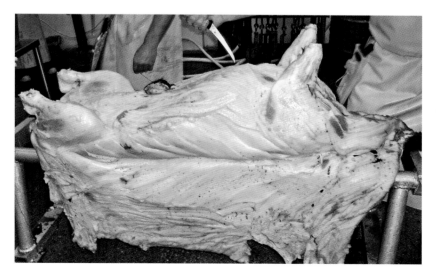

The breastbone can be severed with a sharp, sturdy knife. Position it at a point ahead of the first pair of ribs. Use the heel of the blade, which is more solid than the tip of the knife.

To lift the carcass for opening the body cavity, push the hooks through the area between the bone and the tendons on the rear legs. The tendons are strong enough to suspend the weight of the carcass.

BUTCHERING AT HOME

Whether you are butchering at home or having a pig processed at a local locker, you will have to decide which pig to process if you have more than one to consider. Pork carcasses for home use should be the highest quality you produce and come from pigs that are from five to eight months of age. Pigs fed liberal amounts of quality feeds grow rapidly and will produce pork cuts of proper size and finish.

Butchering at home requires some thought before, during, and after the entire process. Planning ahead for this event will minimize mistakes, reduce the chance of injury to helpers, and provide you with quality meat products for your family.

DEVELOP A PLAN

Develop a list of all the steps needed, from beginning to end, several days before butchering your pig. Planning ahead will reduce any surprises during the butchering process and help organize the event in a logical, efficient manner. Having a list of the equipment and butchering tools required will ensure everything is at hand when needed. Attention must be paid to cleanliness at all times

so the meat isn't contaminated during any part of the process. A review of the skeletal structure, digestive system and placement of organs, and some understanding of the circulatory system will be useful when you are cutting the carcass. Knowing the precise location of the jugular veins in the neck will help you make a clean, swift kill.

The location where you plan to carry out the kill should be properly equipped for the job. A shed or building that is free from dust or outside elements can provide a good place for the initial stages. If this is used, construct a small holding pen near the butchering table to reduce the distance the carcass has to be carried.

There are two ways to accomplish the removal of the skin and hair. One is by manually skinning the pig and the other by using a scalding procedure. Manually skinning a pig requires less effort and equipment to accomplish the same task. If you choose to use the scalding method, you will need a convenient heating arrangement, such as a scalding vat, and an efficient way of swinging the carcass into the boiling water with a block and tackle or some other apparatus. Pig carcasses can be hung from sturdy tree limbs, heavy gambrel sticks, or metal meat hooks. Whichever method you choose will require enough strength and support to hold a large pig upside down and be near the scalding vat and cooling tubs.

A proper set of butchering tools includes sticking knives, skinning knives, boning knives, butcher knives, steel sharpener, meat saws, and meat hooks. Other useful items include thermometers, meat grinder, meat needles for sewing rolled cuts, hair scrapers, hand washtubs, clean dry towels, soap, and vats for hot and cold water.

Providing a proper location and sharp tools will aid in more efficient slaughtering and less time spent looking for items at critical moments.

CARE BEFORE BUTCHERING

You should take the attitude that your butchering process begins several days before the actual event. Confine the pig to be butchered to a small solitary pen two or three days prior to the butchering. Provide plenty of fresh water, but significantly restrict feed twenty-four hours before butchering so the pig has less material in its stomach and intestines. Providing a cool and calm environment several days beforehand will keep the pig rested and quiet. Never attempt to butcher a pig that is overheated, excited, or fatigued. When the body temperature is above normal the meat easily becomes feverish and is difficult to chill properly; poorly chilled meat cannot be properly cured. This increase in temperature can cause the meat to sour or be tainted before it is cut up.

Some spoilage and low quality meat can be directly attributed to natural forms of bacteria that have been allowed to develop and multiply. The bacteria found in the blood and tissues of a live pig must be held in check and prevented from multiplying until the meat is cured. This is one reason butchering was historically done in the early spring or late fall of the year, when the weather was cool. Think of it as a race between the bacterial action in the blood and tissues that want to multiply and the curing agents used such as salt, cold water, and other factors that depress bacterial growth. You need to win this race.

DISPATCHING YOUR PIG

When the butchering tools are laid out; the table is thoroughly washed with soapy water, rinsed, and dried; enough help is on hand; dripping pans are ready to catch blood; ice is in the cooling vat; and everything is in place, you can dispatch your pig.

This is perhaps the most critical time in the whole process, and if you feel uncomfortable sticking a knife into the throat of a pig, you may want to have someone else adept at the task handle this part of the process. This should be arranged prior to this day and not be a last-minute decision.

Butchering a pig by only sticking it with a knife is the most practical, efficient, and humane method of killing a pig. Other methods such as shooting are less reliable unless the pig is agitated. A good bleed is difficult to obtain when the pig is shot.

You can stick the pig either in a raised or prone position, depending on whether or not you have equipment to raise it in the air. To hoist the pig in the air, a chain or strap can be looped between the hock and the hoof in order not to bruise the hams. The pig has less ability to free itself if hanging upside down than if it is rolled on its back and the feet held by several people while one person sticks it. This upside down posture tends to immobilize the pig and makes the actions with your knife easier and safer for you. The most satisfactory bleed occurs when a pig's head hangs downward. Be aware that this position will be very uncomfortable for the pig, and it will typically flail its feet that are free. The feet must be firmly immobilized, as they can be used by the pig as effective weapons of defense.

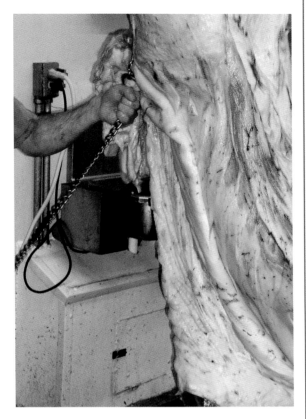

After the carcass is raised, you can remove the rest of the hide by pulling downward. This should be sufficient to fully strip it from the carcass.

To open the body cavity, slice down and out with the heel of the knife. Be careful not to cut or sever the intestines. Do not rush this part of the process, as it avoids any fecal contamination of the carcass that might occur if the intestines are cut.

Once the belly wall has been opened, the internal organs and viscera will spill outward but will still be held by the diaphragm and gullet. Pull the organs and viscera forward to aid in cutting the final attachments.

A 240-pound pig will typically have about 30 to 40 pounds of viscera and organs to handle. They can be dropped into a clean tub positioned below the carcass or lifted and carried to a clean table.

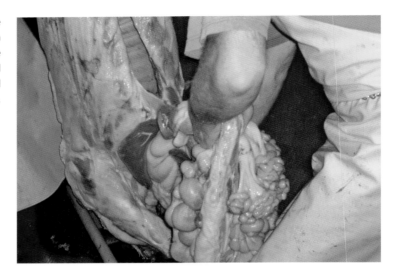

The final cut will divide or sever the diaphragm and open the rest of the body cavity to allow the final drainage of blood from the carcass.

With the skin and internal organs now removed, the final step in this part of the process is to split the carcass in half and wash it with clean water. The carcass is now ready for cooling before being cut into the primal portion.

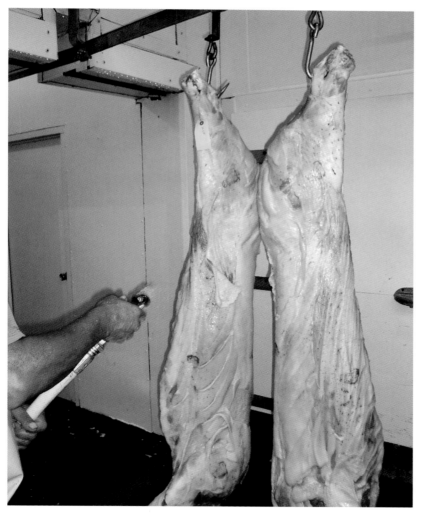

The carcass should hang in a cooler for about seven to ten days to allow it to cool and age. Chilling the carcass makes trimming the fat and cutting the meat much easier.

Pork sausage links are made from casings which are filled with ground meat and seasonings such as salt, pepper, and sage. The long, filled casings are twisted and shaped into links. These can be smoked and prepared for eating by pan frying, roasting, baking, or braising.

The process of breaking down a carcass involves first disconnecting the end portions. These can be further broken down into smaller, more individual cuts. Begin by taking one half of the carcass—a side—and cutting at either end. Use a meat saw to remove the shoulder, front leg, and head by counting two ribs out from the neck. Saw until the shoulder is disconnected from the loin and ribs.

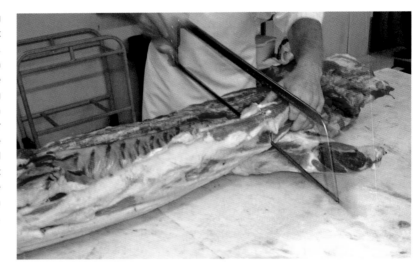

The ham and leg shank can be set aside for trimming and deboning. A typical ham, such as the one shown at right, can weigh between 16 to 18 pounds.

The upper portion of the shoulder is the Boston butt, which lies above the picnic ham. The division between these approximates the horizontal line of fat in the center of this section.

Fat that has been left on the carcass during the chill process is more easily removed from a cold carcass than a warm one.

When the pig is safely immobilized, press the sharp blade edge of the sticking knife in front of the point of the breast bone and quickly slide it in to make a short vertical incision about four inches long in the center of the neck. This should sever the jugular vein and release large quantities of blood. Having tubs placed below to catch the blood will make cleaning up easier and allow you to work around the pig in dry conditions. The knife should not be inserted so far into the neck that it enters the chest cavity, as this may cause internal bleeding and blood clots.

Do not stick the heart, as it is needed to continue working properly to pump out the blood as rapidly as possible. Cutting the heart will cause internal bleeding and create lower-quality meat. The key to this phase is to get a good bleed as quickly as possible. When the blood flow has stopped or slowed to a drip, your pig can be moved to a table where it can be skinned, which requires less time and effort than if scalding it.

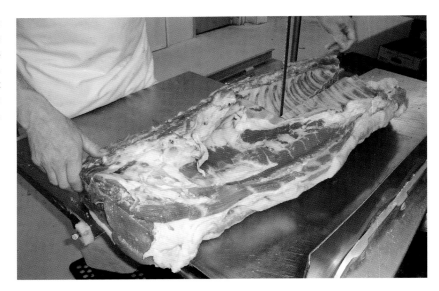

The third primal cut can remove the top ribs near the backbone. This disconnects the belly from the loin, but be careful not to cut into the tenderloin, which lies just in front of it.

SKINNING

Skinning is the removal of the hide from the pig and takes off the outside layer without using hot water or the extra effort of scraping the hair. Most home butchering will not require the use of the skin, which is generally discarded. In the past the skin was left on the bacon and hams to protect them, but with modern refrigeration this is not required.

Skinning a pig includes removing the skin from the belly without puncturing the abdomen with your knife. This is best accomplished by laying the carcass on a table or trolley or suspending it at an appropriate work height.

With a short skinning knife, begin your first cuts at the rear ankles and slice completely around them, but avoid cutting the tendons above the hocks. Cut down the inside center of each leg to a point below the pelvis, and avoid deep cuts into the meat. Do the same with the front legs, and cut to a center point at the base of the chest. Use your knife to score a line down the center of the belly, from the anus to the base of the chest, without penetrating the abdominal wall. Start at the chest and create tension on the skin by pulling it away with one hand as you slice with the other. This tension will help separate the skin from the body.

When finished with both front legs, start on the belly and slowly work to the rear, pulling the skin away from the center until you reach the base. Do the same with the other side. Start at the top of the rear legs by pulling the skin over the hams. At this point the skin should be loose from the belly and legs, and by pulling downward and slicing the skin, the weight of the skin will create tension to help with the rest of the process. Once the skin is completely removed, it can be set aside.

SCALDING

Scalding the carcass is one way to remove the hair on the skin, although this process requires more time and effort to accomplish the same thing as skinning the carcass.

However, for those who wish to use this procedure, proper equipment makes the job easier and you can use a tank or barrel for this step. The tank needs to be filled with water brought to a boil prior to sticking your pig in it. The water can be boiled by using a pit fire underneath the tank, a gas or propane heater, or other means to safely raise the water temperature. This saves time in keeping the process moving, because it is difficult to raise the water temperature once the pig is immersed in the water. The water should be kept between 150 and 160 degrees Fahrenheit. If you are using a long horizontal tank, rotate the carcass until the hair starts to slip. If you are using a barrel, first lower the head into the water while the feet and legs are dry. Then turn it around and place the meat hooks in the lower jaw and lower the rear end into the boiling water. Using an accurate thermometer will help you maintain the temperature to make the scalding easier and eliminate the chance of the hair setting tight against the skin. After lifting the carcass from the scalding water, wash it clean with hot water, scrape off any remaining scruff, and then rinse it down again with cold water.

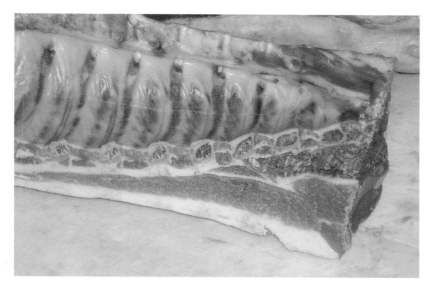

The pork loin is cut out as a whole section and then the various cuts can be made from it. The whole pork loin will include the top loin chop, rib chop, loin chop, sirloin chop, and baby back ribs.

Baby back ribs lie from the blade to the center section of the loin and are trimmed out after the tenderloin is removed. They generally consist of at least eight ribs and are typically sold by the pound.

Disconnect the belly from the loin area by using a meat saw or band saw to keep your cut straight. Do not cut into the loin area or you will damage a valuable cut.

Use the saw and cut just below the bulge at the base of the ham. This will separate the ham from the pork shank. You can trim the fat from the ham at this point or do it after you've made all the other cuts.

The sirloin end roast is cut from the hip bone to the center cut loin. It is about four inches wide and makes a wonderful roast. Try to square it up for even cooking.

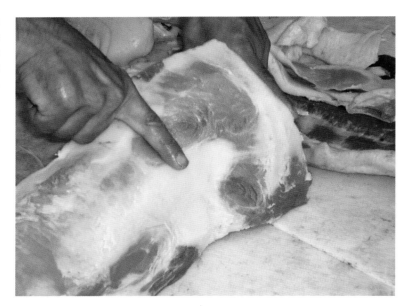

Scraping

Scraping the hair off the skin is the next step of the butchering process. Some of it can be done while the pig is still in the scalding vat, when you are ready to lift the carcass out and place it on the table. Using a scraper, start at the head and feet, as these areas are the first to cool. Your scraping strokes should go in the direction the hair lies, as it will come off easier. After the hair has been removed, use your scraper in a circular motion to work out dirt or scruff that may be embedded in the skin. A soft bristle brush is useful for cleaning up the carcass once the hair is removed. Any stray bristles of hair can be removed with a little hot water and a sharp knife.

HANGING

If the pig is laid out on a table, locate the area between the foot and the hock on the rear legs. Make deep cuts up the center of the bone on each leg to find three tendons. Use your fingers to pull the tendons out and slip the gambrel stick through one and then the other. These tendons are strong enough to hold a hanging carcass while you open up the body cavity. Before you make any incisions to open the carcass, be sure all knives and butchering tools have been scalded and cleaned. Any knife or other tool to be used should be scalded again before use if they have been dropped on the floor. From here on, cleanliness is absolutely essential.

The four ribs in front of the sirloin end roast are cut for country-style ribs. This portion lies just above the tenderloin and is part of the pig's backbone.

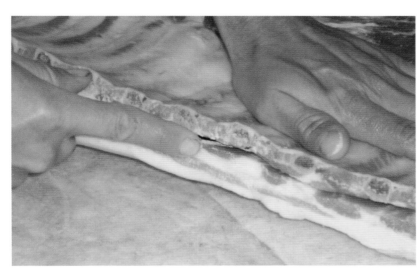

The spare ribs are that part of the rib cage below the backbone down through the belly and lie just behind the diaphragm. Having sliced these with the meat saw, you can slice them from the belly. Trim as close to the bone as possible.

REMOVE THE HEAD FIRST

Removing the head first accomplishes two things: it gets it out of the way and aids in quickly cooling the carcass. Also, removing the head permits blood to be completely drained from the carcass. Begin by cutting above the ears at the first joint of the backbone and then across the back of the neck. When you reach the windpipe and throat, cut through them and the head will drop, but don't slice off the head completely just yet. Pull down on the ears and continue your cut around the ears to the eyes and then toward the point of the jaw bone. When you slice through the last part of the skin at the end of the jaw, the head will come free, but the jowls will still be attached. Wash the head quickly and trim it as soon as possible.

SPLITTING THE CARCASS

Splitting the carcass is easiest to accomplish when it is suspended. Cut a line down the center of the belly between the hams to the sticking point at the base of the chest, but do not cut through the belly wall. To split the breast bone, place the heel of your knife against the bone and cut outward. You may have to work the blade to split the breast bone and divide the first pair of ribs. If your knife will not cut through the breast bone, you may need to use a saw.

In either case, avoid cutting past the upper portion of the breast bone and into the stomach. This is a thin area, and you do not want to cut the stomach open. By cutting through the breast bone and first rib, you will open the chest cavity sufficiently to allow any blood that has accumulated to drain out.

After splitting the breast bone, make an incision near the top of the abdominal wall to pull the skin outward with your fingers. Gravity will pull the intestines down toward the bottom of the chest cavity and leave room for you to insert the knife and your free hand. Grip the handle of the knife with the blade turned toward you. You will slice downward and cut with the heel of the blade and push the intestines away from the knife as you slice down the belly. Keep the intestines away from the blade so you don't cut them and spill their contents, which can contaminate the cavity. As you reach the parted breast bone, the intestines will fall forward and downward. They are still attached by muscle fiber and will not fall far. This is an easier method than drawing the knife upward to slice open the belly. Although it is a bit awkward, it minimizes the chance of puncturing the intestines.

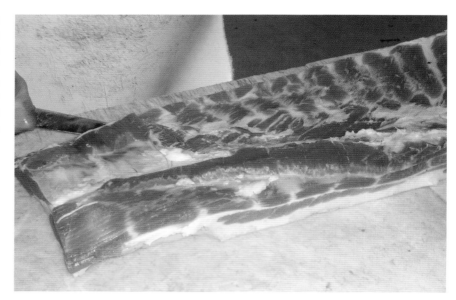

The belly is left for the bacon. Square up the sides and ends by making parallel cuts, and trim off as much fat as possible. This portion can be cut to any thickness, depending on your preference.

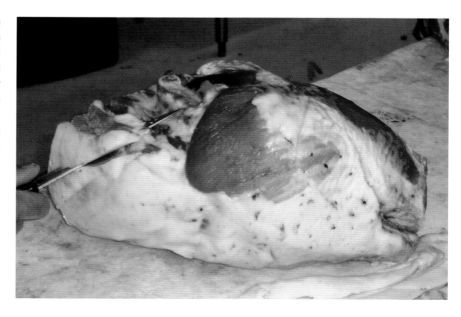

If the ham remains, trim the fat as much as you wish. Cut out the tail bone and trim around the area to remove any unwanted bone. The ham can be smoked, sugar cured, or cut as ham steaks.

The shoulder can be cut into the picnic roast, which is the lower portion and includes the shank, and into the Boston butt, which is the upper portion.

The next step is to split the aitch bone. This will separate the hips and make the cut down the spine easier. First, make a cut in the center between the two hams until you reach the aitch bone. This can be severed either with the heel of the blade or with a meat saw. At this point, the intestines are still suspended by the gut leading to the anus. Before you make a cut around this, tie the end securely with a cord to keep any fecal contents from falling out. Start in front and cut completely around the anus until it is free. This should allow the entrails to fall outward and downward. A tub should be placed under the carcass to catch the entrails as you pull the kidneys, heart, liver, and stomach toward the opening. The diaphragm will now be exposed and you will see the gullet that leads to the stomach. When you sever the gullet, the entire mass of entrails should come free and drop into the tub.

You can place the entrails on a table to cut off the liver and wash it in clean, cold water. Trim out the gall bladder and remove the spleen. The stomach should be tied off with a cord and cut free. The heart and lungs should still be inside the carcass cavity at this point and are located in front of the diaphragm. Make an incision in the diaphragm where the red muscle joins the connective tissue. This will expose the heart and lungs, which should be pulled downward and cut free from the backbone. Trim any fat off the heart and lungs and wash them with cold water.

If you plan to use the intestines for sausage casing, they will have to be cleaned and rinsed with a salt solution several times. The easiest way to accomplish this is to turn them inside out after cutting them into several lengths and scrape off the mucous coating. Generally the small intestines are used for sausages, so tie off the large intestine, sever it from the small intestines, and discard it. The heart and liver can be saved and used to feed pets if you choose.

SPLITTING THE BACKBONE
The hanging carcass needs to be split apart while it is still warm. First, wash the inside of the carcass cavity and, using a meat saw, slice down the center of the backbone. Be sure to make a straight cut or you may damage some of the loin areas. You can leave about 12 inches of skin uncut at the shoulders to keep the carcass from separating if you are concerned about it slipping off the gambrel. If you are not concerned about it slipping, continue to separate the back.

If you choose, the hams can be partially filleted when the carcass is suspended. Start your cut at the flank and continue to follow the curvature of the ham until you reach the pelvis, and then do the same procedure on the other side.

CHILLING THE CARCASS
Your carcass is now ready to be chilled. A cold carcass is easier to trim and cut up than a warm one. Cooling it quickly also minimizes bacterial growth and souring of the meat. It is easier to cool the carcass when it has been split apart as the air circulates around more of the body.

Square up the shape of the shoulder roast for even cooking. This can be further broken down into several smaller portions if desired or it can be left as a large cut.

The clear plate is located at the top of the shoulder and contains little usable meat. First trim this bone off the top of the Boston butt. The meat you trim off the clear plate can be used as part of the mix for making pork sausage.

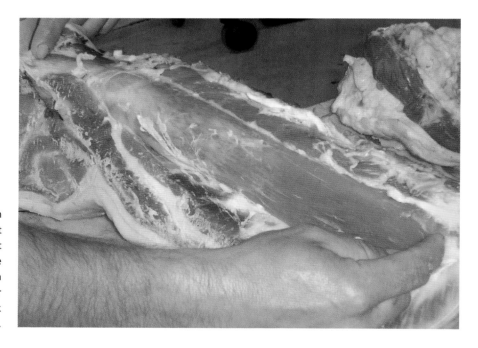

The whole tenderloin is the leanest and most tender cut of pork. It is located along the bottom of the loin starting at the center cut and running back to the sirloin.

To properly chill your carcass, have a separate tub or vat large enough to completely submerse both halves in ice water maintained at a temperature between 34 and 38 degrees Fahrenheit for a minimum of 24 hours. If you are using a refrigerator it is possible to maintain a temperature of 38 degrees Fahrenheit at the bone within 12 to 24 hours. Chilled carcasses should not be worked with until all the tissue heat is gone. When it is thoroughly chilled, you are ready to cut up the carcass.

If you are only interested in cutting up the carcass, and not the butchering process itself, you may be able to arrange a slaughtering date with a local meat processor and receive a chilled carcass to work with at home.

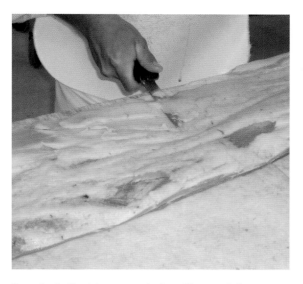

Turn the belly slab over and trim off as much fat as possible. The belly can be sliced thin or thick for different sized bacon strips.

The large roasts can be further broken down into smaller portions. This will make them easier to cook, and they will cook more evenly in smaller, square portions.

The chops can be cut in various thicknesses, depending on your taste and customer orders. They can be packaged individually or in multiple cuts and are generally sold by the pound.

CUTTING THE PORK CARCASS

Before you begin to cut up the carcass, make sure you have sharp knives and several tubs available with cold water mixed with salt to start the curing process. One cup salt for every two gallons of water is a good mixture. There are many ways of creating salt brines for curing meat. Too little salt may cause spoilage, while too much salt makes hard, dry, over-salted meat. Check with your local county or university extension family living educator for assistance.

If you use a meat saw to make the cuts, you should scrape the bone dust from the cuts after sawing. This mixture of small particles of meat and bone results from sawing. Cleaning off the cuts or the saw blades makes them less "crunchy" and reduces the chances of creating a bacterial hostel.

Place one side of the carcass on a clean table and start by removing the ham. Using the meat saw, cut off the leg bone at the hock and cut through the fifth and sixth lumbar vertebrae. You can also make a cut through the stifle joint, which is located between the hock and the base of the ham or ham shank. The ham can be trimmed of fat before it is placed into a salt brine of cold water.

The next step is to saw off the shoulder at the third rib, counting from the neck. The shoulder can be kept whole, cured, smoked, or it can be divided into Boston butt and picnic ham. You can divide the shoulder into picnic ham and Boston butt by cutting about one inch below the shoulder blade and parallel with the breast. Square the picnic by sawing off the foreleg. The foreleg will contain some red meat and is more useful than the rear leg, which contains little meat.

Sausage can be processed as single links or in labeled packages. There are several packaging options for the products you wish to market. A local licensed meat processor may be able to help you decide which option best suits your needs.

Bacon can be vacuum sealed and kept refrigerated to retain its flavor and freshness without the use of preservatives. Pork products can still have an acceptable shelf life if kept frozen or refrigerated below 40 degrees Fahrenheit at all times.

To separate the loin, make a straight cut from a point close to the lower edge of the backbone at the shoulder to a point just below the tenderloin muscle from which the ham was cut. This will separate the loin from the belly or side. When trimming the loin, leave about ¼ inch of fat. The belly or side can be sliced into strips for bacon. Trim off excess fat and the pork belly is ready to be cured or smoked.

These divisions will separate the carcass into the five major areas that can be further divided into smaller portions. Manuals are available to help with creating the finer cuts from a carcass. Contact your local county agricultural extension office for further assistance. They can often direct you to additional information or to a local licensed meat processor who may be able to provide a slaughtering service at a future date.

Although it may be possible for you to accomplish the entire butchering process yourself or with help, it may be in your best interests to discuss your needs with a local licensed meat processing facility. Many have reasonably priced services that can quickly and easily process your pig. Your county agricultural extension office should be able to provide you with more information about the facilities in your area.

CHAPTER 8
COUNTRY LIVING, GETTING HELP, AND EXIT STRATEGIES

If you choose to move to the country to pursue a pig-raising enterprise— or if you already live in the country and decide to start raising pigs— you'll face some big changes. Key to the success of your venture is getting along with your neighbors. There are steps you can take to make the transition smooth.

You can find help with this and other issues through an array of agencies, schools, and networks.

You might also want to plan ahead for the possibility of selling your farm. An exit strategy is good to have, and it makes sense to think about it as you plan the other aspects of your business.

LIVING IN THE COUNTRY

Country living offers many benefits to those who wish to move away from the urban lifestyle, and those benefits may hold a special attraction for you. If you have already moved into a rural area, you should be aware of these same considerations, although you may view them in a different way.

The purpose of your farm may be production agriculture, but the purpose of a neighbor's rural home might be just to enjoy all the benefits of living in the wide open spaces without having to contend with many urban issues, including traffic, close neighbors, and noise at night.

With two different agendas at work in the same area, it's understandable that conflicts and disagreements arise as to which agenda takes precedence. If conflict arises between farmers and new rural neighbors, it's usually over such things as odors from hauling manure, dust from working fields in the spring, or noise (especially if harvesting is done at night). There are other things that can cause grievances between these two groups of people, but these three seem to be the most common.

There are connections between those living in a rural area, even if the parties concerned fail to recognize them at first. The dynamics between different agendas influence the quality of life in any rural area. As a producer you can have a positive impact by acknowledging both agendas. The cumulative effect over time will enrich your life and the lives of your neighbors.

Many states now give farmers a basic right to farm without the fear of lawsuits brought by offended neighbors. In these cases an agricultural operation is presumed not to be a nuisance to the neighbors even when new neighbors move in. If the farm operations are conducted in a reasonable manner, the new neighbors legally can't complain. Landowners, residents, and visitors must be prepared to accept the effects of agriculture and rural living as normal and understand they are

Open rural areas attract urbanites to the countryside because of the perceived peace and solitude that exists, and it may be similar to what draws you there. The movement of people from villages and cities to the countryside increases the potential for problems and misunderstandings. However, it does not have to lead to them. Opening discussions with new neighbors may help defuse confrontations when they may not understand typical farming practices.

likely to encounter a number of these practices as area farmers follow their normal farming routine.

However, there are different conditions attached to these ideas in some states, because the right-to-farm laws do not give farmers complete freedom to do as they please. In these instances farmers must operate in a legal and reasonable manner to be eligible for the law's protection. Some states have developed a list of specific annoyances that are not considered a legal nuisance to neighbors, including odor, noise, dust, and the use of pesticides or other chemicals, the very conditions which, without the laws, could lead to a lawsuit by a neighbor.

IN YOUR BACKYARD

It is often impossible to stop people from purchasing land near or adjacent to your farm, and you have little control over what their expectations are for moving into the neighborhood. You may feel they should be grateful for having the chance to live where you do, just as you may have decided to venture into farming for similar reasons. But that may not be the position they take, and they may feel it is their right to be there without the inconveniences agricultural production imposes upon them. You may find yourself confronted with disenchanted

neighbors over some slight misunderstanding, perceived or otherwise, but fortunately there are ways of disarming any confrontational situation before it gets out of hand. Taking a proactive approach before it reaches this stage may be your best defense.

BUILDING BRIDGES

Many conflicts can be solved, and some avoided altogether, by using strategies to build stronger ties with your nonfarming neighbors and your local community. One key is trying to get producers and consumers talking the same language. Communication is very important. Perhaps the best way to avoid conflict is to use responsible farm management practices that contribute to the best environment for everyone. Protection of groundwater supplies, controlling odors when possible, and keeping weeds from becoming a problem are some practices that can highlight your commitment to a good relationship. Besides being good stewardship practices, these will be your allies if you have a disagreeable neighbor who may prefer a legal confrontation.

Unexpected new neighbors are an important reason for having good fences around the perimeter of your farm. It is not acceptable to let your animals wander and roam onto another's property, assuming that since you were there first, you do not need to maintain your fences. Similarly,

Having disposable plastic boots available for visitors is an easy way to create a barrier against the potential introduction of disease on your farm from outside sources. While these may seem excessive precautions to some, they can help maintain a healthy herd.

the fences are a visual boundary for your neighbor and a reminder they must also respect your property and not allow their animals to roam on your land without permission.

Talking with your new neighbors is another way of diminishing potential problems. Simple gestures like giving them a farm tour, hosting picnics, giving away free manure, or providing garden space in a corner of a field that is not farmed could be a way of reaching across the divide. The purpose is to create a stronger sense of community, which may be one of the reasons that brought you to your farm in the first place. Don't lose sight of that attraction, because it can have a similar hold on other people.

FARM VISITS

One step you can initiate is to invite your new neighbors to visit your farm and see it close at hand. A friendly atmosphere where you can discuss your farming practices, procedures, the goals of your business, the values you hold for the land, and the reasons for involving your family in the farm operation may help offset their concerns as they witness your family's total involvement.

Familiarizing your neighbors with the reasons you chose to live in the country and work with pigs may give them a better sense that sometimes allowances will need to be made on their part so that your farming practices can continue unimpeded. Explaining how things work on your farm may be helpful if your new neighbors have little or no knowledge about the seasonal requirements of farming. There may be times when field work is done at odd hours of the night or day. Helping your neighbors understand that these hours occur sporadically and usually only during planting and harvesting may ease their concerns.

You may find that one reason your neighbors moved to a rural area is for the health of a family member. Developing a sense of being neighbors and knowing of conditions where you may be able to prevent distress within their family is a good way of avoiding conflict. By knowing these conditions, it will be easier for you to inform them of your intentions for field work or harvesting near their home several days ahead of time so that they can make adjustments to their schedules and routines and avoid the area where you will be working.

Although you may not feel that it is your position to make the first contact, etiquette and politeness are always in fashion, especially toward

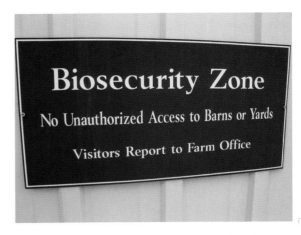

In recent years more efforts have been made to keep infections and viruses from being introduced onto farms from outside sources. Placing new animals in quarantine while conducting blood tests, physical examinations, and visual inspections are tools that can be used to create and maintain a healthy pig herd.

your neighbors. Working through conflict involves finding common ground and shared interests. In this case, the shared interest is the desire of farmers and nonfarmers, whether new or long-time residents, to enjoy the kind of life that the countryside has to offer, the economic benefits that can be derived, and the sense of community that can be developed. As farm families continue to have more new neighbors, building bridges will be vital to working through these challenges. Never underestimate the power of communication to avoid or solve problems before they arrive. When a problem or conflict does present itself, you can feel confident that a workable solution can be made.

BIOSECURITY CONCERNS

The communication and personal contact with your neighbors must be balanced with the need for biosecurity on your farm, regardless of its size. In the context of livestock production, biosecurity refers to measures taken to keep disease agents out of your swine herd where they do not already exist. These measures exist on three levels: national, state, and individual herds.

Biosecurity should not be confused with bioterrorism or agro-terrorism, because they are not the same thing, although biosecurity is a part of each concern. As a farm owner, the responsibility for the biosecurity on your farm rests with you, as well as the plans and steps taken to prevent the introduction of any infectious disease onto your farm or to limit the spread of any disease already present within your herd. To be successful, your plans must address how infected pigs will be isolated from the herd and how cleaning and disinfecting procedures will be used.

The greatest risk of introducing an infectious disease onto your farm is to bring in new pigs that have been exposed to a disease or have it themselves. This is one reason that visual observation of any new pig is important as the first line of defense with the initial screening.

Insisting upon and making use of certain diagnostic tests can prevent problems from occurring where none exist. A discussion with your local veterinarian about isolation procedures for newly purchased pigs or pigs already infected in your herd will give you a plan to implement before it is needed. Your veterinarian can also help develop a vaccination program and disinfection plan that will raise your herd's resistance to infectious diseases.

Your concern over biosecurity does not need to impair your relationship with your neighbors. Most established farmers are already aware of these concerns, and some may have procedures already in place. Common sense should help dictate your approach to biosecurity as well. Having visitors is a good way to expose them to your farm, but asking them to wear protective foot, boot, or shoe coverage will subtly show that you are serious about the health of your pigs. It would be the rare visitor who refuses your request to put on a pair of disposable plastic slipover shoe covers before entering your barn or farmyard.

DEFENDING FARM PRACTICES

Animal rights campaigns occasionally reach front page news and are a subject all farmers must take time to understand. By giving humane care to their animals, farmers for generations have known that it is in their best interests and the comfort and thriving nature of their animals to provide them with certain unspoken rights such as food, shelter, light, sufficient space, and clean barns.

This stewardship has been appropriated in the last few decades by groups whose agenda has included the elimination of animal agriculture. Considering that only 1.6 percent of the United States population lives on farms today, there is no question that agriculture is getting smaller and urbanization is growing.

Marked waste containers can be used for disposal of plastic boots after your visitors leave. Using plastic boots helps prevent visitors from contaminating your farm or inadvertently taking contaminants away with them.

In the urban population this has led to a gulf between perception and reality of how animals are treated, and this has much to do with urbanization itself. Urban people generally experience animals differently than farmers. To them animals are viewed as house pets and treated, in many cases, as another member of the family.

Animal rights have become a social issue rather than a scientific one. As these views spread and gain traction, it is up to livestock producers to address these perceptions and defend their practices not just scientifically but ethically. It is about what the animals represent and the values associated with raising them.

Training kits are available to help you and your family understand the issues and how to respond to those who do not. Contact your county agricultural extension office for assistance. Raising animals that provide food for other people is noble work and one that has given great satisfaction to many farm families.

HELP AND ADVICE

You may have questions from time to time about how to handle certain problems that arise on your farm. Help is available to assist you in getting answers to your questions, solutions to your problems, and new perspectives to consider.

EXTENSION SERVICES

The agricultural extension service, usually located at your county seat, can provide help with answers and solutions. Extension educators are specially trained and have access to their state's university systems and research departments. They are able to glean a large amount of information from these sources and pass it on to you.

Because of their extensive contacts across the country and increasingly around the world, extension educators receive information on traditional practices as well as the latest innovations to arrive on the scene as new approaches make front page news in farm publications. Although you may have the same access to electronic data, their background and experience can help narrow your focus, and their services are free.

TECHNICAL SCHOOLS

Technical schools provide information, training, and assistance by offering a hands-on approach. Classes offered by technical schools can supplement and provide training that may not be available through county agricultural extension offices. Their classes may help you prepare other areas of your farming operation not directly related to livestock, such as accounting or financial assessments.

Counties or regions within many states have grazing networks that hold informal monthly meetings, with pasture walks as the featured agenda. These are not only learning sessions where ideas can be exchanged, but also social events that can lend support and encouragement to your efforts.

Advice and assistance can come from many different sources, including established and reputable breeders, county extension offices, universities, and state and federal agencies. There is a vast amount of information available to help you with your questions and pig-raising program.

GRAZING NETWORKS

Grazing networks are an excellent way to get to know other farmers in your area or region who are developing grazing programs. These are informal groups whose purpose is to help beginning grazers solve problems, explore alternative methods, provide social contacts, and generally have fun while learning about the way other grazers handle their pastures, problems, and successes. Because these networks are made up of like-minded people, ranging from experienced to novice, the network goals are usually to help one another rather than force views upon new members.

Grazing networks typically have a mix of livestock producers in their membership, with dairy, beef, pig, and sheep owners among the most prevalent. The main focus of the grazing network is the pasture walk that typically takes place during the main grazing season, from April to October. Typically pasture walks are held on a different farm

each month and begin with a brief introduction by the host farmer or extension agent with a short discussion prior to going into a pasture to walk and observe the conditions the host farm is using. Questions are encouraged. Topics covered can include pasture composition or what grasses are planted and being used for forage, fencing layout, how that particular system works for the owner, watering systems, lanes, and the overall grazing management being used. Some networks use a series of walks on the same farm during the year to follow the farm's progress and learn how the host farmer handles the changes over the grazing season.

UNIVERSITY SYSTEMS

All land-grant colleges in the United States provide agricultural classes that are available to the public. The university system can provide research data on a variety of subjects. These reports may help in making decisions related to your farming practices.

There are many organizations and breed associations that may be useful to your program by providing low cost or no cost information and assistance. There are also institutes or associations that aim to create environmentally and economically sustainable rural communities and regions through sound agriculture and trade policies.

At one time the independent status of university research was unquestioned. However, the increasing budget constraints of universities have caused them to seek funding in some cases through private companies and foundations. The private nature of these companies and foundations has caused some concern about the ability of the universities to remain unbiased when the finances are provided by companies with a profit motive. The research may not be tainted, but whether the wrong questions are being asked or the right questions are not being asked or researched may be a result of losing financial support and should make you closely examine new studies being promoted.

STATE DEPARTMENTS OF AGRICULTURE
The United States Department of Agriculture (USDA) is the cabinet-level agency that oversees the vast national agricultural sector. Its duties range

from research to food safety to land stewardship. Every state in the country has a department of agriculture that administers the programs of that particular state and operates under its statutes. These agencies have a wide range of booklets, pamphlets, and other publications available to help you understand the rules and laws, obtain licenses, and comply with regulations pertaining to your farming operation.

ONLINE RESOURCES
The resources available to you are limited only by the amount of time you dedicate to researching them. The Internet has made available huge quantities of information that may apply to your situation. As with all things related to electronic data, it is best to check the sources of the articles you use, and to use care in sharing sensitive information about your farm.

EXIT STRATEGIES

Change happens in life and on farms. For anyone living on a farm there comes a time when a sale occurs, whether a complete sale precipitating an exit from farming or a partial sale of the assets, such as the pigs or machinery, and retention of the land. It may be a sale to another family member or a combination of these instances. As with any other aspect of your business, planning ahead for a sale will generally pay greater dividends than if you let events dictate your course of action.

The reality is that a sale is likely to occur in your lifetime. There is no disgrace in this situation because farms have been bought and sold for generations as people have entered and exited farming. That may be the cold reality, but there is little doubt that leaving a farm can be an emotional time for you and your family. The reasons for leaving may include health issues, finances, changes in a family situation, or a desire to pursue other interests.

In many cases it may seem easier to get into a business than to get out. If you plan to exit from your swine operation, there are some basic similarities and differences to any other business that has an exit strategy. A discussion with your financial advisor may provide answers to help you successfully exit farming.

DON'T FEEL GUILTY

You do not have to feel guilty that you are leaving the farm or having a sale. Too often families who leave their farm develop a sense of guilt because they consider it a failure on their part. Perhaps financial problems contributed to the sale, but the fact that they spent time and effort in trying to

The one constant in life is change—much like the seasons. By having a clear purpose at the beginning of your program, you may be able to accommodate change and use it to your advantage. If you decide to exit your pig-raising program, you still may have options for using your farm. Changes in your circumstances may require alternative plans, but a well-thought-out program at the beginning can help decrease the anxiety of making those decisions.

make their business succeed should not be viewed as a failure. Farming can be a challenge even in the best of times. Market forces have a way of blowing down the best-built house, and those who can withstand such events are sometimes more lucky than good managers.

SALE OPTIONS

Farms are different from many other businesses in that you have living assets—animals—that must be sold. They are similar to other businesses because there are also nonliving assets to be sold and tax considerations after a sale. There are several ways to handle the sale of your pigs or machinery, and each has advantages and disadvantages. A thorough understanding of each option can save you money and minimize surprises.

If you have few pigs to sell, your choice of exit may be simply to market them through a packing house if they are of appropriate weight. Or you may try to sell them at public auction or privately to another pork producer. Exiting a pig-raising program is a relatively simple procedure, but ending it does not necessarily mean an end to your farming enterprise.

The conditions for having your sale may determine what happens next on your farm. If you sold all the animals from your farm, then the land will still be part of your assets. Your facilities can stay empty if the pigs are gone, but this will mean

you will have to pay taxes on unused buildings. The buildings can be rented to another party to provide income. A written contract for renting to another party is good business and will keep any potential problems to a minimum. Additionally you can rent out your land and still retain ownership and live on the farm. Another option for your farm if you discontinue raising pigs is to convert your land into a vegetable produce farm or other cash crops.

POSTSCRIPT

From these examples it should be obvious that the sale of your pigs does not need to be the end of your farming life, unless you choose it to be. There have been many farmers who enter a swine operation, leave, come back in, and then leave again. Some try to avoid the peaks and valleys of pork marketing cycles and use an entry and exit policy.

Another reason for selling and going back into pork production may have to do with wanting time away from the farm or health and personal issues. Selling your pigs does not have to be a traumatic experience if you accept the idea that a sale will happen eventually either with or without you in attendance. Planning ahead for a sale can take a lot of the emotional stress out of the decision. It can leave you with a healthier state of mind and a satisfaction that you have accomplished the goals you set at the time you entered the pig-raising business.

At the end of the day, selling your pigs does not mean you need to leave the farm, unless you choose to. Some producers use an entrance and exit policy for many reasons, including health issues, finances, changes in a family situation, or to pursue other interests.

APPENDIX 1
GLOSSARY

abortion: Loss of swine fetuses between breeding and up to the 109th day of pregnancy.

aitch bone: The rump bone.

barrow: A male pig that has been castrated, usually as a young pig.

boar: Adult male pig kept for breeding purposes.

bred: Has mated and is pregnant.

breed: To mate. Also refers to a group of animals with common ancestors and physical characteristics.

bulky feed: A feed that is usually higher in fiber and lower in energy.

carcass: The dressed body of an animal slaughtered for food.

castration: To remove the testicles of the male pig.

confinement: To keep pigs close together without allowing them outside access.

creep feed: A feed given to young pigs from one week of age to weaning.

crossbred: A pig whose parents are of different breeds.

dam: Female parent.

dry sow: A sow that has weaned her pigs and is no longer producing milk. Also refers to a pregnant sow any time prior to farrowing.

estrus (heat cycle): The regular intervals between heat periods, usually nineteen to twenty-one days with an average female.

farrow: To give birth.

farrowing stall: A narrow stall where the sow farrows, usually 20 to 24 inches wide and 7 to 8 feet long, used to keep sows from crushing their young.

feed efficiency: The number of pounds of feed required by an animal to gain 1 pound in body weight.

feeder pig: A young pig that has been weaned and is now ready to feed out, usually between 25 to 40 pounds.

feeder pig sale: A place where feeder pigs are sorted into uniform groups by weight and grade and sold.

finish: Feeding a pig to market weight.

foster: Practice of placing piglets from mothers with too many piglets to adequately feed to mothers with extra udder space; usually occurs in first few days of birth.

free choice: Providing pigs with grain and a protein supplement in a self feeder.

full feed: Giving a pig all it will eat.

gestation period: The time between the mating of a sow and birth of the pigs, about 114 days in swine.

gilt: A young female pig that has never farrowed.

heat: Time that a fertile female can be bred; also describes the behavior of a female at the time she will accept the male.

heat lamp: A 250-watt lamp suspended about 24 inches above newborn pigs to keep them warm.

hog: A generic term that can be applied to a growing or mature pig.

lactating period: The period of time the sow produces milk and the piglets are nursing.

length of hog: A measurement taken on the carcass or estimated on the live hog. It is the distance from the front edge of the first rib to the front end of the aitch bone.

litter: Offspring produced at one farrowing.

loineye area: The number of square inches in a cross-section of loineye muscle. This is taken by cutting the loin between the tenth and eleventh ribs and measuring the cut surface area of the muscle.

meat-type hog: A pig that exhibits a high lean-to-fat body ratio.

palatable: A feed that tastes good to the pig and is easily digestible.

pasture: An area or field that consists of permanent or annual grasses, clover, alfalfa, or other legumes where pigs can freely graze.

pig crushing: When the mother inadvertently lies on her piglets or several pigs lie on top of each other to stay warm. Both events usually involve the death of one or more animals.

piglet: A newborn pig of either sex, about 2 to 3 pounds and usually less than eight weeks old.

purebred: Both sire and dam are of the same breed.

rations: The diets formulated for pigs and other livestock.

runt: A small, undersized, or weak piglet in the litter.

shoat, shote: A pig of either sex after weaning; usually weighs less than 100 pounds.

sire: Male parent.

sow: Female pig over one year of age that has farrowed a litter of pigs; rhymes with cow.

wean: To remove piglets from their mother; take off milk.

APPENDIX 2
RESOURCES

BREED ASSOCIATIONS:

American Berkshire Association
2637 Yeager Road
W. Lafayette, IN 47906
(765) 497-3618
www.americanberkshire.com

American Landrace Association
Member of the National Swine Registry
2639 Yeager Road
W. Lafayette, IN 47906
(765) 463-3594
www.nationalswine.com

Note: Yes, these two organizations (above) are just down the road from each other.

American Mulefoot Hog Association and Registry
18995 V Drive
Tekonsha, MI 49092
(517) 518-7930
www.mulefootpigs.tripod.com

American Yorkshire Club
Member of National Swine Registry
2639 Yeager Road
W. Lafayette, IN 47906
(765) 463-3594
www.nationalswine.com

Chester White Swine Registry
Member of Certified Pedigreed Swine (CPS)
P.O. Box 9758
Peoria, IL 61612
(309) 691-0151
www.cpsswine.com

Hampshire Swine Registry
Member of National Swine Registry
2639 Yeager Road
W. Lafayette, IN 47906
(765) 463-3594
www.nationalswine.com

Large Black Hog Association
C/O Felicia Krock, Registrar
(800) 687-1942
www.largeblackhogassociation.org
Note: No postal address available.

National Hereford Hog Record Association
C/O Becky Hyett, Secretary
826 140th Street
Aledo, IL 61231
(309) 299-5122
www.nationalherefordhogassociation.com

National Spotted Swine Record, Inc.
Member of Certified Pedigreed Swine (CPS)
P.O. Box 9758
Peoria, IL 61612
(309) 693-1804
www.cpsswine.com

Poland China Record Association
Member of Certified Pedigreed Swine (CPS)
P.O. Box 9758
Peoria, IL 61612
(309) 693-1804
www.cpsswine.com

Red Wattle Hog Association
C/O Kathy Bottorff
41 Jones Road
Horse Cave, KY 42749
(270) 565-3815
www.redwattlehogassociation.com

Tamworth Swine Association
C/O Shirley Brattan, Secretary
621 N CR 850 W
Greencastle, IN 46135
(765) 653-4913
www.tanworthswine.org

United Duroc Swine Registry
Member of National Swine Registry
2639 Yeager Road
W. Lafayette, IN 47906
(765) 463-3594
www.nationalswine.com

OTHER ORGANIZATIONS

The Livestock Conservancy
P.O. Box 477
Pittsboro, NC 27312
(919) 542-5704
www.livestockconservancy.org

National 4-H Council
7100 Connecticut Avenue
Chevy Chase, MD 20815
(301) 961-2800
www.4-h.org

National Association of Animal Breeders (NAAB)
P.O. Box 1033
Columbia, MO 65205
(573) 445-4406

National FFA Organization
P.O. Box 68960, 6060 FFA Drive
Indianapolis, IN 46268
(317) 802-6060
www.ffa.org

National Pork Producers Council
www.nppc.org
Washington office:
122 C Street, NW
Suite 875
Washington, D.C. 20001
(202) 347-3600
Des Moines office:
10664 Justin Dr
Urbandale, IA 50322
(515) 278-8012

INDEX

Animal identification, 37

Behavior, understanding, 111–112
Biosecurity concerns, 162–163
Breeding systems, 109. *See also* Reproduction
Breeds
    Associations for, 25–26
    Berkshire, 14
    Chester White, 14–15, 25, 27
    Choctaw, 24
    Choosing, 12
    Duroc, 15–16, 25
    Guinea Hog, 24
    Hampshire, 16–17, 25, 28
    Hereford hog, 20–21
    Heritage, 20–25
    Landrace, 13–14, 25
    Large Black, 21
    Mulefoot, 22
    Ossabaw Island Hog, 24
    Poland China, 17–18, 25
    Red Wattle, 22–23
    Spotted Poland China, 18–19, 25
    Tamworth, 23–24
    Variations in, 13
    Yorkshire, 19, 25, 27
Butchering
    Chilling carcass, 155–156
    Cuts of meat, 141–143
    Cutting carcass, 157–159
    Dispatching pig, 145, 149
    Hanging, 152
    At home, 140
    Plan for, 144
    Preparations for, 144–145
    Process of, 145–149
    Removing head, 153
    Scalding, 150
    Scraping, 152

    Skinning, 150
    Splitting backbone, 155
    Splitting carcass, 153–155
Buying pigs, 32–36

Certified Pedigreed Swine (CPS), 25
Conventional farming, 43
Country living, 160–163
County agricultural extension services, 164
Cultural images, 6–7

Environmental Assurance, 96
Environmental issues, 96
Euthanasia, 118
Exit strategies, 167–168
Extension services, 164

Farm location, 31
Farm visits, 162
Farming practices, 38–43
    Defending, 163
Farrowing, 44–45, 52, 55, 56–57, 102–103, 105–106
Feed
    Additives for, 84
    Consumption, 82
    Storage, 83
    See also Nutrition
Fencing
    Barbed-wire, 72
    Boundary lines and, 62–63
    Cable-wire, 73
    Construction of, 67–71, 76–77
    Electric, 72–73
    High-tensile, 73
    Materials for, 67, 69
    Permanent, 65–67
    Planning, 63–65
    Posts for, 74–76
    Temporary, 65–67

    Tools for, 74–76
    Wire panels for, 74
    Wood, 70–72
    Woven-wire, 70

Grazing networks, 165

Health
    Alternative treatments, 120–123
    Common disease, 113–116
    Concerns, 37
    Conventional treatments, 118–120
    Disease prevention, 116–117
    Immunization, 117
    Management, 117–123
    Parasite control, 117
    Regulations, 37
    Sanitation, 116–117
Health regulations, 37
History of pigs, 7–9
Housing. *See* Shelter design

Management
    Behavior, understanding, 111–112
    Moving and sorting, 112–113
    Overview of, 111
    Recordkeeping, 123–125
Manure
    Composting, 90, 93
    Handling of, 91–92
    Managing, 88, 95–96
    Pasture, 95
    Pollution and, 89, 91
    Soil conservation and, 96
    Storage options for, 93–94
    Uses for, 89
Marketing
    Advertising and, 138
    Classes for, 131–132

Feeder pigs and, 132–136
Labels and, 137–138
Market hogs and, 132–136
Options for, 129
Plan for, 129
Transportation and, 137
Value-added products and,
  130
Moving techniques, 112–113

National Pork Producers Council
  (NPPC), 96
National Swine Registry (NSR), 25
Nutrients
Sources for, 78–80
Value of, 78
Nutrition
Energy requirements, 80–81
Forages, 84–87
Formulating rations, 86, 87
Minerals, 82, 84
Post-service, 103
Protein requirements, 80–81
Reproduction and, 103
Vitamins, 82, 84
See also Feed; Water

Online resources, 166
Organic farming, 38, 39–41

Pastures
Health and, 49–50
Housing, 52
Overview of, 46–48
Permanent, 48–49
Setting up paddocks, 51–52
Small producers and, 52
Stocking rates and, 50–51
Temporary, 48–49
Pork
Changing tastes and, 9–10
Health benefits of, 10–11

Pork-production systems
Farrow-to wean, 44–45
Farrow-to-finish, 45
Pasture raising and grazing,
  46–48
Weaning-to-finish, 45–46
Profitability, 4–6, 78
Purchasing pigs, 32–36

Registry associations, 25–26
Reproduction
Artificial insemination,
  106–108
Breeding systems and, 109
Estrus, 99–100
Gestation table, 99
Heat detection, 101–102
Lactation, 102–103
Managing boar and, 103–104
Nutrition and, 103
Overview of, 98
Performance, 102–103
Pregnancy rates, 100–101
When to farrow, 105–106

Selling, timing of, 131
Shelter design
Farrowing and, 56–57
Feeder pigs and, 57–58
Overview of, 55–56
Pasture structures and, 59–61
Weaning and, 57–58
Showing, 126–127
Social considerations, 31
Soil conservation, 96
Sorting techniques, 112–113
Start-up economics, 33
State departments of agriculture,
  166
Sustainable farming, 42–43

Technical schools, 164
Transportation, 36–37

University systems, 165–166

Water, 60, 61, 82, 84. See also
  Nutrition

ABOUT THE AUTHOR

Philip Hasheider is a fifth-generation farmer who grew up in southcentral Wisconsin, where he was involved in 4-H and FFA. In recent years he has combined his interests in agriculture and history to write thirteen books, including six for Voyageur Press: *How to Raise Cattle, How to Raise Pigs, How to Raise Sheep, The Family Cow Handbook, The Complete Book of Butchering, Smoking, Curing, and Sausage Making,* and *The Hunter's Guide to Butchering, Smoking & Curing Wild Game & Fish,* all available at qbookshop.com.

ACKNOWLEDGMENTS

I wish to thank my wife, Mary, for offering her insights, support, critiques, and encouragement; each has made this book better.

Our son, Marcus, willingly spent many hours with me taking most of the photographs that appear in this book.

Our daughter, Julia, with her spontaneous laughter and great sense of humor, was always a support.

My mother, Shirley, has preserved several photographs that appear in this book. With a mother's infinite charity, she has always found her son's writings to be supremely interesting. My brother Neal provided helpful commentary on the text, based on his many years of experience with the United States Department of Agriculture. My brother Bruce helped clarify several important points regarding today's purebred pork industry.

I wish to thank Pam Ziegler for allowing the use of my nephew David's cover photograph. In one brief moment she captured his enthusiasm, passion, and love for pigs.

A special thank you to Jeff and Chris Sorg for agreeing to my repeated requests of their time and patient assistance.

There were others whose assistance was essential to the completion of this book, including: Paul Dietmann; Rich Lange; Tom and Diane Rake; Tony and Sue Renger; John and Jim Straka; Jack, Sharon, Jonathan, Joel, and Beth Wyttenbach; and Gary and Rosemary Zimmer.

Several people helped secure photographs and information about the heritage breeds of pigs they raise, including Shirley Brattain; Dave and Jill LaFollette; Ruby Schrecengost; Ted and Gayle Smith; and Josh and Kelly Wendland. I appreciate the time and assistance they willingly gave to me.

I particularly wish to thank Jerry and Ruth Apps, Beverly Davidson, Petrina Green, and Dr. Helen Hotz for supporting my writing efforts and offering continued encouragement.

Also, a thank you goes to my editor, Amy Glaser, who, once again, provided me with a unique opportunity to present this information to a larger audience.